Jens H. Flamm

Diffraction and Scattering in Launchers of Quasi-Optical Mode Converters for Gyrotrons

Karlsruher Forschungsberichte aus dem
Institut für Hochleistungsimpuls- und Mikrowellentechnik

Herausgeber: Prof. Dr.-Ing. John Jelonnek

Band 2

Diffraction and Scattering in Launchers of Quasi-Optical Mode Converters for Gyrotrons

by
Jens H. Flamm

Dissertation, Karlsruher Institut für Technologie
Fakultät für Elektrotechnik und Informationstechnik, 2012

Impressum

Karlsruher Institut für Technologie (KIT)
KIT Scientific Publishing
Straße am Forum 2
D-76131 Karlsruhe
www.ksp.kit.edu

KIT – Universität des Landes Baden-Württemberg und nationales
Forschungszentrum in der Helmholtz-Gemeinschaft

KIT Scientific Publishing 2012
Print on Demand

ISSN: 2192-2764
ISBN: 978-3-86644-822-3

Foreword of the Editor

Gyrotron oscillators (gyrotrons) are high-power millimeter-wave electron tubes mainly used for electron cyclotron resonance heating (ECRH) and plasma stabilization (current drive (ECCD)) through localized non-inductive current drive in magnetically confined plasmas for energy generation by controlled thermonuclear fusion. Gyrotrons are electron cyclotron resonance masers operating in a longitudinal magnetic field configuration, usually produced by a superconducting magnet. 2 MW gyrotrons with coaxial cavities for improved mode selection are under development for the worldwide fusion experiment - International Thermonuclear Experimental Reactor (ITER) - at Cadarache, France. Quasi-optical mode converters are key elements of high-power gyrotrons. Located after the interaction cavity the generated RF-beam is separated from the electron beam and the higher-order cavity mode is transformed into a fundamental Gaussian field distribution. The quasi-optical system consists of the waveguide antenna (so-called launcher) followed by typically three quasi-optical mirrors. The launcher is a tapered hollow waveguide with specific inner wall perturbations. The launcher is transforming the higher-order cavity mode into a mode mixture which is radiated as linearly polarized paraxial beam. The mirrors are further transforming and guiding the RF-beam. The quality of the quasi-optical system is described by the Gaussian beam content and the stray radiation of the launcher inside the mirror box.

Dr.-Ing. Jens Flamm is comparing, ranking and extending different methods for calculation of diffraction and scattering in launchers (waveguide antennas). Four methods are considered, three scalar methods and one method based on the Electric Field Integral Equation (EFIE). It is shown how to reduce the computation time of the scalar methods using numerical techniques applied from digital signal processing. Consideration of diffraction inside the launcher is reducing the problem size. Comparison and ranking of the different methods allows estimation of the design accuracy and calculation error. Together with the achieved enhancement in numerical efficiency it allows analysis of production tolerances and opens the possibility to synthesis of tapered launchers for single-frequency and future multi-frequency gyrotrons with adaptive surface perturbation in acceptable time. The newly introduced depolarization factor helps to estimate stray radiation in the future.

Dr.-Ing. Jens Flamm is providing the gyrotron community an essential tool for future design of quasi-optical mode converters. It will help to analyze and to design advanced gyrotrons for even higher output power as well as for multi-frequency operation. We are wishing Dr.-Ing. Jens Flamm much of further success in his professional career. We are sure that his future activities will benefit from his excellent abilities, his knowledge gained and his tools derived within these studies.

Diffraction and Scattering in Launchers of Quasi-Optical Mode Converters for Gyrotrons

Zur Erlangung des akademischen Grades eines

DOKTOR-INGENIEURS

von der Fakultät für
Elektrotechnik und Informationstechnik
des Karlsruher Institut für Technologie (KIT)

genehmigte

DISSERTATION

von

Dipl.-Ing. Jens Hanspeter Flamm

geb. in Karlsruhe

Tag der mündlichen Prüfung: 03.02.2012

Hauptreferent: Prof. Dr. rer. nat. Dr. h.c. Manfred Thumm
Korreferent: Prof. Dr.-Ing. Dr. h.c. Klaus Schünemann

Kurzfassung

In der vorliegenden Arbeit werden unterschiedliche Methoden zur Berechnung von Beugung und Streuung in Launchern (Hohlleiterantennen) von quasi-optischen Wellentypwandlern in Gyrotrons verglichen, klassifiziert und erweitert. Die Erweiterung dient der Berücksichtigung einer mittleren Radiusänderung der Hohlleiterantenne. Der Vergleich und die Erweiterung der schnellen Berechnungsmethoden zur Komponentensynthese ermöglicht die Reduktion der Beugung und Streustrahlung von zuverlässigen und leistungsfähigen Millimeterwellenquellen für die Heizung und Stabilisierung von Kernfusionsplasmen.

Zunächst wurde ein Vergleich von drei skalaren und eines vektoriellen Berechnungsmodells zur Berechnung der elektromagnetischen Wellenausbreitung in Launchern erstellt. Im Rahmen des Vergleichs wurden zwei Methoden mit einem erstmalig angewandten Algorithmus aus der Digitalen Signalverarbeitung implementiert. Unter Berücksichtigung der Beugung innerhalb eines Launchers konnte die Problemgröße und somit die Rechenzeit mit diesem Algorithmus um den Faktor fünf, im Vergleich zu einer bestehenden Implementierung, reduziert werden. Der Vergleich der vier Berechnungsmodelle ergab praktisch identische Werte für skalare und vektorielle Korrelationskoeffizienten der abgestrahlten Feldgröße. Der Geschwindigkeitsgewinn durch die neuartige Implementierung eröffnet die Möglichkeiten zur Fertigungstoleranzuntersuchungen wie auch zur Berechnung von Gyrotronlaunchern von Mehrfrequenzgyrotrons mit adaptiven Oberflächenstörungen in annehmbarem Zeitrahmen.

Desweiteren wurde eine adaptive Version des eingeführten Algorithmus eingesetzt, um somit erstmalig die Möglichkeit der schnellen skalaren Feldberechnung in getaperten Gyrotronlaunchern mit adaptiven Oberflächenstörungen zu geben. Dieser neu angewendete Algorithmus wurde durch ein kommerziell verfügbares 3D Feldberechnungsprogramm verifiziert, welches um den Faktor 1800 langsamer ist. Mit der Einbettung dieses Algorithmus in einen Oberflächensynthesealgorithmus eröffnet sich erstmalig die Möglichkeit direkt getaperte Gyrotronlauncher mit adaptiver Oberflächenstörung, zur Reduktion der Streustrahlung, zu synthetisieren.

Zusätzlich kann der, im Anhang der Arbeit, eingeführte Depolarisationsfaktor zur zukünftigen Abschätzung der erzeugten Streustrahlung in Gyrotronlaunchern herangezogen werden.

Abstract

In this work different methods for the calculation of diffraction and scattering in launchers (waveguide antennas) of quasi-optical mode converters for gyrotrons are compared, ranked and extended. The extension gives the opportunity to take a tapered average radius of the waveguide antenna into account. The comparison and the extension of the fast field calculation methods for component synthesis opens the possibility to reduce diffraction and stray radiation of reliable and powerful millimeter wave sources for heating and current drive in magnetically confined plasmas.

The comparison consisted of four methods calculating the electromagnetic wave propagation in launchers. Of these four methods, three are scalar quasi-optical methods and one is a vectorial method based on the Electric Field Integral Equation (EFIE). In the framework of the comparison two methods have been implemented with an algorithm from digital signal processing for the first time. Consideration of diffraction inside the launcher reduced the problem size and together with the new algorithm the computational time was therefore reduced by a factor five, compared to the existing implementation. The comparison of the four methods gave practically identical results for the scalar and vector correlation coefficients of the fields. The gain in computational efficiency opens the possibility to investigate production tolerances as well as calculation of gyrotron launchers of multi frequency gyrotrons with adaptive surface perturbation in an acceptable time frame.

Furthermore the adapted version of the newly introduced algorithm gives the possibility for fast scalar field calculation in tapered average radius gyrotron launchers with adaptive surface perturbation for the first time. This new applied algorithm was verified by comparison to the commercially available 3D field solver. The computational time of the scalar method is reduced by a factor of 1800 compared to the 3D field solver. Embedding this new algorithm into a launcher surface synthesis procedure opens the possibility to directly synthesize tapered gyrotron launchers with adaptive surface perturbation and therefore reduce the stray radiation. In addition the introduced depolarization factor, found in the appendix, helps to estimate this stray radiation in the future.

Contents

List of used Symbols and Variables

$\vec{A}$	magnetic vector potential
A_i, A_j	mode amplitude of mode i or mode j
A_{1-3}, B_{1-3}	constants
A_{mn}, B_{mn}	coefficients
$a_{\Delta m}(z)$	amplitude of wall deformation in the waveguide
$\vec{B}, B_0$	magnetic flux density vector, magnetic flux density
$\vec{C}, \vec{K}$	arbitrary vectors
c_0	speed of light
$\vec{D}$	electric flux density vector
$\vec{E}$	electric field vector
$e^{r.m.s}$	residual error (root mean square)
e	Euler's number, charge of an electron
$\vec{e}$	unit vector
$\vec{F}$	electric vector potential
f	frequency of an electromagnetic field or wave
G	Green's function
g_m	amplitude of cylindrical wave
$\vec{H}$	magnetic field vector
$H_m^{(n)}$	Hankelfunctions of order m and kind n
J_m	Bessel function of first kind and order m
J_m'	derivative of the Bessel function of first kind and order m
$\vec{J}$	current density vector
j	imaginary unit $\sqrt{-1}$
K_{ij}	coupling coefficients between mode i and mode j
$\vec{k}$	free space propagation vector

k, k_z, k_ρ	free space wave number, wave number in z-direction, wave number in radial direction
$\Delta k_{z}, \Delta m$	difference in axial wave number
$k_\perp$	wave number in transverse direction
L	length of the launcher
L_B, L_φ	Brillouin length, ray length along the circumference
L_{trans}, L_C	ray length transverse in the (ρ,φ)-plane, length of launcher cut
Δm	waveguide perturbation order, difference in azimuthal order
m_e	rest mass of the electron
m, n	mode index: azimuthal, radial
N_m	Neumann function of order m
$\vec{n}$	normal vector
P	point in space
R, Φ, Z	functions
R_0, R_c	waveguide radius, caustic radius
$\Delta R(\varphi, z)$	surface perturbation of launcher wall
$\vec{R}$	ray vector
$\vec{r}_0, \vec{r}$	distance vector, vector
r_0, r, d	distance
ρ, φ, z	cylindrical coordinates
S, dS	surface, surface element
s	harmonic factor of Ω_c
t	time, parameter
u, ψ	scalar field, satisfying the Helmholtz equation
$\psi^{(1)}, \psi^{(2)}$	inward, outward traveling part of the scalar field
V_{acc}	accelerating voltage
V, dV	volume, volume element
$v, v_\parallel, v_\perp$	velocity, axial, transversal
x, y, z	cartesian coordinates
α	slope angle, parameter
$\eta_{vector}, \eta_{scalar}$	vector and scalar correlation coefficients

ε, μ	dielectric permittivity, magnetic permeability
φ_m	magnetic scalar potential
γ	relativistic factor (Lorentz factor)
λ	wave length
Ω_c	electron-cyclotron angular frequency
ω	angular frequency of an electromagnetic field or wave
Ψ	Brillouin angle
χ_{mn}	n-th root of the Bessel function of order m
χ'_{mn}	n-th root of the derivative of the Bessel function of order m
$\Delta\chi'_{mn}$	difference in Bessel roots
$\vartheta(\varphi, z)$	phase correction
ξ, ζ	functions
∇	Nabla operator
θ	Spread angle
$\mathcal{F}, \mathcal{F}^{-1}$	Fourier Transform, inverse Fourier Transform
$\mathcal{L}, \mathcal{T}$	mathematical operators

1 Introduction

In order to satisfy today's rapidly growing demand for energy, alternatives to power plants using fossil fuels or nuclear fission need to be found. Among the promising candidates are power plants using thermonuclear fusion in magnetically confined plasmas. The advantages of fusion based power plants are practically everlasting fuel resources and environmental reasons, such as no need for final disposal of nuclear fuel rods, the partial recycling of the fuel in use and no CO_2 production. Thermonuclear fusion is achieved by heating a gas mixture, typically consisting of Deuterium and Tritium, up to several million degrees centigrade to form a plasma. Such high temperatures are necessary to overcome the Coulomb barrier of the two nuclei, initially repelling each other, and make the fusion possible. The produced neutrons from the fusion process transfer the energy in form of heat to a Lithium blanket, from which a turbine can be operated. In addition to the energy transfer, the neutrons together with Lithium are used to breed the necessary Tritium fuel. The generated α-particles keep the plasma temperature high.

An efficient method to heat up the plasma to the necessary temperature is heating with microwaves. This method is called Electron-Cyclotron-Resonance-Heating (ECRH). The ECRH method heats the electrons in the plasma through the electron-cyclotron resonance effect. The electrons then transfer the energy to the ions via collisions. The advantages of this method over other employed methods, such as Neutral Beam Injection or Ion Cyclotron Resonance Heating, are the small sizes of the antennas, the very high coupling factor into the plasma and the possibility of localized heating. Furthermore the microwaves can be used to control and stabilize the plasma through local non-inductive current drive (ECCD). The minimal overall power level necessary for start-up of the plasma in future fusion reactors is $50-100$ MW and the necessary frequency, due to the magnetic confinement, for the energy transfer to the electrons lies between 100 and 250 GHz. In today's plasma experiments power levels of some 10 MW are employed. For control and stabilization of the plasma some 10 MW are necessary. Currently the only feasible coherent microwave source in this frequency and power regime for ECRH is the gyrotron (gyro-oscillator or gyromonotron). State-of-the-art gyrotrons with a conventional cylindrical cavity are able to produce above 1 MW continuous wave (CW) of microwave output power at frequencies of 110–170 GHz. In pulsed operation, currently for CW operation developed, gyrotrons with a coaxial-cavity are capable of producing over 2 MW RF output power at 170 GHz. A current overview of state-of-the-art gyrotrons, gyro-devices and Free Electron Lasers and their fields of application is given in [Thu11].

1.1 Gyrotron

The gyrotron is an electron vacuum tube, that covers mainly the millimeter and sub-millimeter wavelength range gap between conventional microwave tubes (e.g. klystrons, magnetrons, etc.) and far-infrared lasers. Newest gyrotron developments extend into the THz-regime. The gyrotron is based on the electron cyclotron maser (microwave amplification by stimulated emission of radiation) instability, discovered independently in the late 1950s by several scientists [Twi58, Sch59, Gap59]. This instability uses the fact that moving electrons gyrate in the presence of an axial magnetic field due to the Lorentz force, hence the name gyro-tron. Figure 1.1 shows a schematic drawing of a state-of-the-art high power gyromonotron with coaxial cavity and lateral RF output. In a gyrotron the following major parts can be distinguished: Electron gun (cathode and anode), beam tunnel, cavity, quasi-optical mode converter, collector, dielectric output window, magnets and, in case of coaxial cavity, a coaxial insert.

Starting from the cathode of the electron gun, a hollow electron beam is emitted from the emitter ring. The electrons in the beam are accelerated by the anode and start to gyrate around the magnetic field lines produced by the coils of the magnet. As a result of the two motions, the electrons form helical trajectories. In the beam tunnel, where the magnetic flux density B_z rises, the transverse velocity $v_\perp$ of the electrons is increased. This results in a transfer of kinetic energy from the axial motion into the transverse motion and is necessary since the interaction of the electrons takes place in the plane transverse to the z-axis. While traveling through the cavity, a cylindrical metallic waveguide, the electrons give part of their transverse energy to the electromagnetic wave, which is excited from the noise floor that is present. Since the interaction mechanism of the cyclotron maser instability works in a plane transverse to the longitudinal direction of the

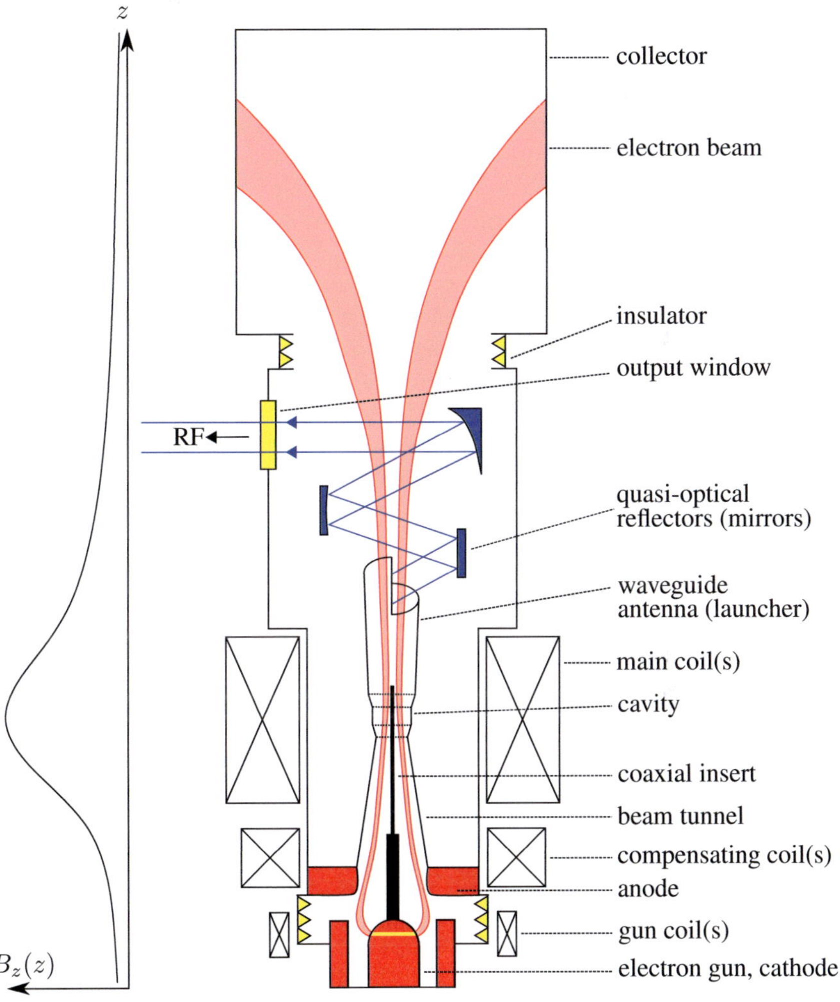

Figure 1.1: Schematic drawing of a high power gyromonotron with coaxial cavity

electrons, the gyrotron is considered a fast-wave device, where the phase velocity of the electromagnetic wave is faster than the speed of light. This fact allows the wavelength to be smaller than the characteristic size of the structure emitting the radiation and leads to a much higher power handling capability in a gyrotron than in conventional tubes. More on the interaction of electrons and electromagnetic wave is given in the subsequent section. The generated electromagnetic wave in the form of a high-order transverse electric mode gets reflected at the lower cut-off section of the cavity and coupled out of the cavity through an output taper at the upper end. Subsequently the wave travels upwards towards the quasi-optical mode converter. This quasi-optical mode converter consists of a waveguide antenna launching a wave beam onto several deflecting mirrors. It is one of the crucial components for continuous wave operation and is further addressed in a subsequent section. The RF wave beam exits the tube through an output window. The spent electron beam travels towards the collector and is swept over the large collector surface to stay within an acceptable range of thermal stress of the collector. The insulator separating the collector and the gyrotron body allows a depressed collector operation to regain some of the residual energy from the electrons.

1.1.1 Gyrotron interaction

The following explanation of the gyrotron interaction gives an overview, since the main focus in this work lies on the waveguide antenna of the quasi-optical mode converter. For an in-depth treatment of the gyrotron interaction mechanism see [Edg93, KBT04, Nus04].
In order to transfer energy from the electrons to an electromagnetic wave in a cavity, the electrons and the wave have to be in phase in some periodic motion. In this case the electron motion can reinforce the electromagnetic wave. The necessary periodic motion is fulfilled by weakly relativistic electrons gyrating in a favorable phase position. The corresponding resonance condition can be derived as follows:
The gyration of the electrons with time can be described using the relativistic cyclotron frequency Ω_c given as:

$$\Omega_c = \frac{eB_0}{\gamma m_e} \approx 2\pi \cdot \frac{28\,\mathrm{GHz} \cdot B_0/T}{\gamma} \tag{1.1}$$

where e denotes the electric charge of the electron, m_e denotes the rest mass of the electron, B_0 denotes the magnetic flux density of the constant magnetic field in the cavity and

$$\gamma = \frac{1}{\sqrt{1 - \left(\frac{v}{c_0}\right)^2}} = 1 + \frac{W_{Kin}}{m_e c_0^2} \approx 1 + \frac{W_{Kin}}{511\,\text{keV}} \tag{1.2}$$

is the relativistic factor, with W_{Kin} denoting the kinetic energy of the accelerated electrons.

For the electromagnetic wave in the waveguide, the following dispersion relation, for the RF frequency ω, can be formulated

$$\omega = c_0 \sqrt{k_\perp^2 + k_z^2} \tag{1.3}$$

with

$$\omega = 2\pi f = k c_0$$

and k denotes the free space wave number of the wave. k_z and $k_\perp$ are the axial and transverse wave number of the transverse-electric (TE) mode in the cavity. Now for the electron and the wave to be in phase over the time t, while traveling in z-direction, the following phase condition has to be satisfied:

$$\frac{d}{dt} \left(\overbrace{\omega t - k_z v_z t}^{\text{e.m. wave}} \overbrace{- s\Omega_c t}^{\text{electrons}} \right) \cong 0 \quad , \quad z = v_z t \tag{1.4}$$

leading to the resonance condition also considered the beam line equation

$$\omega - k_z v_z \cong s\Omega_c \tag{1.5}$$

$k_z v_z$ is considered the Doppler term, due to the relative motion of the electrons to the wave field, consisting of the axial wave number k_z of the wave and the translational drift velocity v_z of the electrons. In addition to the fundamental cyclotron frequency, resonance can also occur at harmonics s of Ω_c. By exciting the electromagnetic wave near cutoff ($k_z \rightarrow 0$), the Doppler term influence becomes very small. When the RF frequency is slightly higher than the cyclotron

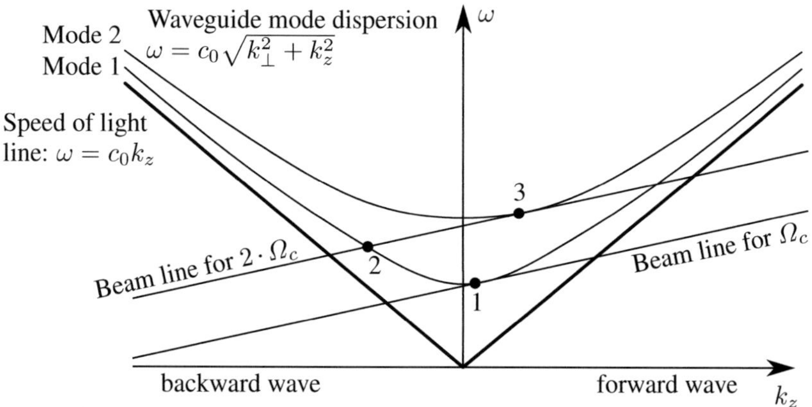

Figure 1.2: Brillouin diagram

frequency, the aforementioned favorable phase position of the electrons is achieved. Without the Doppler term, the frequency of the wave ω is solely determined by the cyclotron frequency and therefore by the magnetic flux density B_0 and not by geometrical constraints. This is the major advantage over conventional tubes. So in order to excite an electromagnetic wave at $f = 170\,\mathrm{GHz}$ with an electron beam with $\gamma \lesssim 1.2$ ($U \approx 90\,\mathrm{kV}$) a magnetic flux of $B_0 \approx 7\,\mathrm{T}$ is necessary.

Figure 1.2 shows a Brillouin diagram, where equations (1.3) and (1.5) are plotted simultaneously, i.e. the wave frequency ω is plotted as a function of the axial wave number k_z. With this diagram different operating points of the gyrotron interaction can be illustrated by selecting points where the curves touch or intersect. Point 1 in figure 1.2 corresponds to the classical gyromonotron interaction of mode 1 near cutoff at the fundamental cyclotron frequency Ω_c. Point 2 shows a backward wave interaction with mode 1, since the corresponding k_z is negative, where the waveguide dispersion hyperbola and the beam line of the second harmonic $2 \cdot \Omega_c$ intersect. Point 3 corresponds to the interaction of a forward traveling mode 2 with the second harmonic of the cyclotron frequency. For numerical calculation of the gyrotron interaction more complicated models have to be considered. The corresponding theories can be found in [Ker96, Jel00].

1.1.2 Quasi-optical mode converter

As explained before, due to a cutoff section at the lower part of the cavity, the generated electromagnetic wave travels upwards towards the quasi-optical mode converter. The purpose of this quasi-optical mode converter is:

- Separation of the generated electromagnetic wave and the spent electron beam

- Extraction of the RF power generated in the cavity by transformation of the complex electromagnetic field configuration into a paraxial wave beam for low loss transmission

This mode converter (Fig. 1.3) consists of a cylindrical waveguide with a helical cut at the end and a mirror system. The waveguide with cut exhibits perturbation of the inner wall and operates like an aperture waveguide antenna. This waveguide antenna is called "launcher". It is used to convert the high-order transverse electric cavity mode into a paraxial beam, reducing diffraction on the edges of the radiating aperture. This beam is then radiated obliquely into free space and deflected by the mirror system to exit the gyrotron horizontally through a dielectric disc window, which is usually made of CVD (chemical vapor deposition) diamond. The surface of the launcher used for the conversion has either a periodic or an adaptive perturbation, depending on the parameters of the TE cavity-mode (Sections 2.2 and 2.3). The goal of the transformation is the adaptation of the electromagnetic wave to free space propagation. The reduction of diffraction at the helical cut results in a reduction of stray radiation inside the tube. Due to thermal reasons, the avoidance of stray radiation is crucial for CW operation of high power gyrotrons [Pri07]. One of the simplest possibilities for a paraxial beam is a Gaussian field profile (TEM_{00}) [Gol98]. The goal is to achieve such a field profile on the radiating aperture of the launcher by employing an appropriate surface perturbation. This generated Gaussian wave beam is then adapted by quasi-optical reflectors (mirrors) to pass through the output window. The first mirror (green like colors) focuses the initially cylindrical wave into a paraxial wave. The second and third mirror (yellow and red like colors) optimize the parameters of the beam to exit the tube with the least window reflection possible.

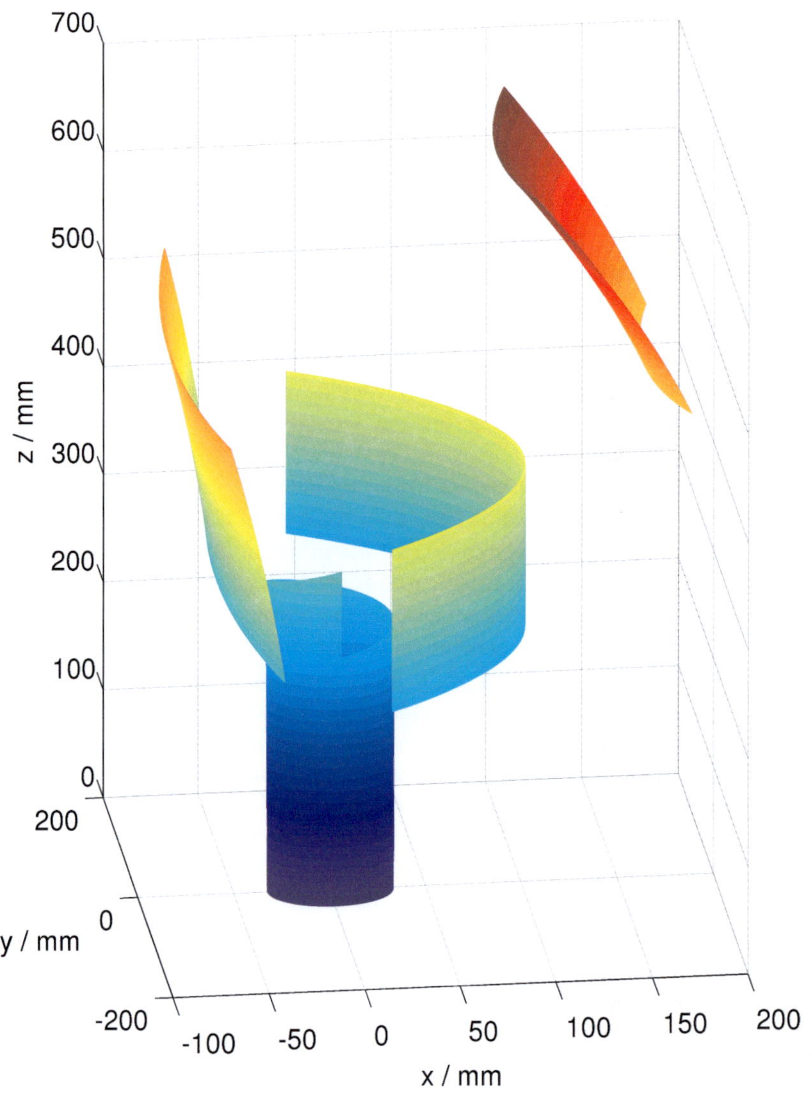

Figure 1.3: Schematic plot of a quasi-optical mode converter for a high power gyrotron

1.2 Goals and structure of the work

The CW operation of a gyrotron will be a essential for heating, controlling and stabilizing power plant scale fusion devices and becomes more critical at higher power levels, e.g. 2 or even 4 MW RF output power [Ber11]. To ensure this operation, the diffraction and stray radiation inside the quasi-optical mode converter have to be kept to a minimum. With higher power levels the necessity for coaxial cavities arises [Ker96]. The operating modes excited in such coaxial cavities have a different structure than the operating modes chosen for a hollow cylindrical cavity when considering the conversion into a paraxial wave beam. The traditional conversion by periodic wall perturbation does not yield an acceptable result [Jin07]. This conversion approach uses the method of coupled modes equations [Pre03, BD04, Sol59] to calculate the conversion of the electromagnetic wave inside the waveguide antenna. By introducing adaptive surface perturbations, a satisfactory conversion of a coaxial cavity operating mode into a paraxial wave beam has been accomplished [JTP$^+$09]. In order to effectively handle the adaptive surface perturbation the slight wall perturbation is modeled as a phase corrector, corresponding to a thin lens in optics, and the wave propagation inside the waveguide antenna is described using scalar diffraction instead of the coupled modes equations. Standard simulation techniques, such as Finite Difference Time Domain (FDTD), are too time and memory consuming for launcher design purposes, since the dimensions of such waveguide antennas are much larger than the wavelength (Diameter $D \approx 20 \sim 30\lambda$ and length $L \approx 100 - 200\lambda$). The reduction to a scalar diffraction problem is possible due to the oversized nature of the problem. Different models of scalar diffraction for these waveguide antennas exist [JTP$^+$09, Kuz97, CDK$^+$06]. In order to classify these methods and evaluate the accuracy of the conversion to the Gaussian beam, the different models are compared to each other and to a commercial full-wave Electric Field Integral Equation solver. Since only one scalar method [JTP$^+$09] and the commercial full-wave code [Nei04] are available at the Institute for Pulsed Power and Microwave Technology (IHM) at the Karlsruhe Institute of Technology, models similar to the two remaining models are implemented in a computer program. In this new approach, the analysis of launchers with tapered average radius is made possible for the first time using the scalar diffraction method. This is accomplished by an adaptation of the segmented convolution algorithm, which is also applied to these problems for the first time. In addition, the segmented convolution offers a considerable speed-up of the calculation. Therefore the possibility to calculate launchers with tapered average radius and adaptive surface perturbation is given for the first time. This taper is necessary to reduce the risk of spurious oscillations inside gyrotron launchers.

This work on the topic of the diffraction and scattering in launchers of quasi-optical mode converters for high power gyrotrons is structured as follows: Chapter 2 derives and explains the required mathematical theory; furthermore it discusses limitations of the theories employed. The subsequent chapter 3 addresses the numerical aspects and algorithms of the four methods used in the comparison as well as the new adaptation of the algorithm for tapered average radius launchers. Chapter 4 applies the introduced methods to an existing un-tapered launcher design for a coaxial cavity gyrotron to verify and compare the newly implemented, as well as the existing, methods. In addition, the verification of the algorithm for tapered average radius launchers, using two example launchers, is given. The final chapter summarizes the work, draws conclusions and gives an outlook addressing future issues and improvements.

2 Waveguide antenna

This chapter is dedicated to the derivation of the required theory, necessary to describe the wave propagation in launchers. It starts with the derivation of the theory of wave propagation in cylindrical waveguides. Then the traditional method for conversion [DPV90] into the Gaussian profile is described. Subsequently three different types of scalar diffraction approaches are discussed as alternatives, describing the wave propagation inside the perturbed cylindrical waveguides in order to overcome the short-comings of the traditional description. The last part of this chapter describes the Electric Field Integral Equation (EFIE), a three dimensional full wave solution, used as the reference solution for the comparison with the introduced scalar propagation models.

2.1 Wave propagation in cylindrical waveguides

2.1.1 Derivation of the wave equation

In order to calculate and describe the wave propagation in a cylindrical waveguide, we start from Maxwell's equations in time harmonic form and derive the wave equation. We choose the time dependence to be $e^{j\omega t} \rightarrow \frac{\partial}{\partial t} = j\omega$. This derivation is similar to those in [Mic98, Dru02]:

$$\nabla \times \vec{E} = -j\omega\vec{B} \tag{2.1}$$

$$\nabla \times \vec{H} = \vec{J} + j\omega\vec{D} \tag{2.2}$$

$$\nabla \cdot \vec{D} = \rho \tag{2.3}$$

$$\nabla \cdot \vec{B} = 0 \tag{2.4}$$

where ∇ denotes the Nabla operator. For linear, isotropic current free ($\vec{J} = \vec{0}$) and solenoidal ($\rho = 0$) media, Maxwell's equations (2.1), (2.2) and (2.3) with the help of $\vec{B} = \mu\vec{H}$ and $\vec{D} = \varepsilon\vec{E}$ become:

$$\nabla \times \vec{E} = -j\omega\mu\vec{H} \tag{2.5}$$

$$\nabla \times \vec{H} = j\omega\varepsilon\vec{E} \tag{2.6}$$

$$\nabla \cdot \varepsilon\vec{E} = 0 \tag{2.7}$$

The solenoidal nature of the field (2.7) lets us introduce an electric vector potential $\vec{F}$ to simplify the calculation:

$$\vec{E} = -\frac{1}{\varepsilon}\,\nabla \times \vec{F} \tag{2.8}$$

Using this relation to replace $\vec{E}$ in equations (2.5) and (2.6), we arrive at

$$-\frac{1}{\varepsilon}\,\nabla \times \nabla \times \vec{F} = -j\omega\mu\vec{H} \tag{2.9}$$

$$\nabla \times \vec{H} = -j\omega\left(\nabla \times \vec{F}\right) \tag{2.10}$$

From equation (2.10) we obtain the following expression for $\vec{H}$:

$$\nabla \times \left(\vec{H} + j\omega\vec{F}\right) = 0 \tag{2.11}$$

$$\vec{H} = -\nabla\varphi_m - j\omega\vec{F} \tag{2.12}$$

In (2.12), an arbitrary scalar potential φ_m was introduced to describe the divergent part of $\vec{F}$ (source-free part of $\vec{H}$), which is not defined by (2.11). Insertion of (2.12) in (2.9) yields

$$-\frac{1}{\varepsilon}\,\nabla \times \nabla \times \vec{F} = -j\omega\mu\left(-\nabla\varphi_m - j\omega\vec{F}\right) \tag{2.13}$$

Using the vector identity [Har61]

$$\nabla \times \nabla \times = \nabla\left(\nabla \cdot\right) - \nabla^2$$

in (2.13) gives

$$-\frac{1}{\varepsilon}\left[\nabla\left(\nabla \cdot \vec{F}\right) - \nabla^2\vec{F}\right] = -j\omega\mu\left(-\nabla\varphi_m - j\omega\vec{F}\right) \tag{2.14}$$

Sorting the equation leads to:

$$\nabla^2 \vec{F} + \omega^2 \mu\varepsilon \vec{F} = j\omega\varepsilon\mu\nabla\left(\varphi_m + \frac{\nabla\cdot\vec{F}}{j\omega\varepsilon\mu}\right) \tag{2.15}$$

Subsequently, choosing the scalar potential φ_m according to the Lorenz gauge[1] as:

$$\frac{\nabla\cdot\vec{F}}{j\omega\mu\varepsilon} = -\varphi_m \tag{2.16}$$

to separate the equations for the vector and scalar potential, we arrive at the wave equation for the electric vector potential $\vec{F}$:

$$\nabla^2 \vec{F} + k^2 \vec{F} = 0 \tag{2.17}$$

In (2.17), the wave number k was introduced as

$$k^2 = \omega^2 \mu\varepsilon. \tag{2.18}$$

Using (2.6) and due to (2.4), a similar derivation is possible for a magnetic vector potential $\vec{A}$ defined as $\vec{B} = \nabla \times \vec{A}$ and results in:

$$\nabla^2 \vec{A} + k^2 \vec{A} = 0 \tag{2.19}$$

With $\vec{F}$ known, one can calculate the electric and magnetic fields using equations (2.8), (2.12) and (2.16):

$$\vec{E}_F = -\frac{1}{\varepsilon}\nabla\times\vec{F} \tag{2.20}$$

$$\vec{H}_F = -j\omega\vec{F} - j\frac{\nabla(\nabla\cdot\vec{F})}{\omega\mu\varepsilon} \tag{2.21}$$

The analog solution for the magnetic vector potential $\vec{A}$ is:

$$\vec{H}_A = \frac{1}{\mu}\nabla\times\vec{A} \tag{2.22}$$

$$\vec{E}_A = -j\omega\vec{A} - j\frac{\nabla(\nabla\cdot\vec{A})}{\omega\mu\varepsilon} \tag{2.23}$$

Here the subscripts "F" and "A" explicitly indicate the derivation from an electric or magnetic vector potential.

---

[1] sometimes mistaken for the Lorentz gauge [NS01]

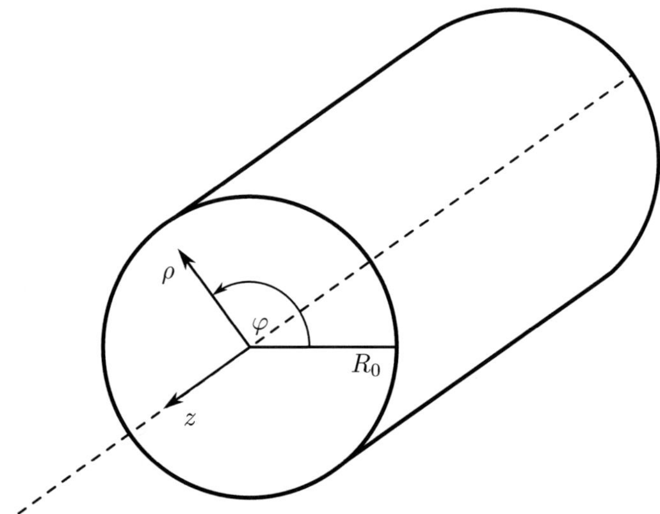

Figure 2.1: Geometry of a circular cylindrical waveguide

2.1.2 Solution for circular cylindrical waveguides

We now consider the solution of either equation (2.17) or equation (2.19) in cylindrical coordinates and unperturbed boundaries, that is for a hollow and smooth metal waveguide with radius R_0 shown in figure 2.1. The resulting boundary conditions at the perfectly conducting waveguide wall are:

$$\vec{n} \times \vec{E} = \vec{0} \tag{2.24}$$

$$\vec{n} \cdot \vec{B} = 0 \tag{2.25}$$

Circular cylindrical waveguides can, as is well known, support different field configurations called modes. The two different supported types are transverse electric (TE) and transverse magnetic (TM) modes. Choosing $\vec{C}$, corresponding to either the magnetic or electric vector potential, solely with a z-component ($\vec{C} = C_z\,\vec{e}_z$)

$$
\begin{aligned}
\vec{K} = \nabla \times \vec{C} \;=\;\; & \vec{e}_\rho \left(\frac{1}{\rho}\frac{\partial C_z}{\partial \varphi} - \frac{\partial C_\varphi}{\partial z} \right) \\
+\;\; & \vec{e}_\varphi \left(\frac{\partial C_\rho}{\partial z} - \frac{\partial C_z}{\partial \rho} \right) \\
+\;\; & \vec{e}_z \left(\frac{1}{\rho}\frac{\partial}{\partial \rho}(\rho\,C_\varphi) - \frac{1}{\rho}\frac{\partial C_\rho}{\partial \varphi} \right)
\end{aligned}
\tag{2.26}
$$

the resulting field, denoted here as $\vec{K}$, has no z-component, as desired for TE-($\vec{E}$) or TM-($\vec{H}$) modes [Bal89].

For TE modes, we choose:

$$
\begin{aligned}
\vec{A} &= \vec{0} \tag{2.27}\\
\vec{F} &= \vec{e}_z \cdot F_z \tag{2.28}
\end{aligned}
$$

and for TM modes the choice is:

$$
\begin{aligned}
\vec{A} &= \vec{e}_z \cdot A_z \tag{2.29}\\
\vec{F} &= \vec{0} \tag{2.30}
\end{aligned}
$$

Now the wave equations (2.17) and (2.19) can be reduced to a scalar equation. For $\psi \equiv F_z \equiv A_z$ the (scalar) Helmholtz equation reads:

$$
\nabla^2\,\psi(\rho, \varphi, z) + k^2\,\psi(\rho, \varphi, z) = 0
\tag{2.31}
$$

Expressing the Nabla operator in cylindrical coordinates gives [Bro05]:

$$
\frac{1}{\rho}\frac{\partial}{\partial \rho}\left(\rho\frac{\partial \psi}{\partial \rho} \right) + \frac{1}{\rho^2}\frac{\partial^2 \psi}{\partial \varphi^2} + \frac{\partial^2 \psi}{\partial z^2} + k^2\,\psi = 0
\tag{2.32}
$$

Equation (2.32) can be solved by the separation of variables of the function ψ, that is we search for a solution of the form $\psi(\rho, \varphi, z) = R(\rho)\cdot\Phi(\varphi)\cdot Z(z)$. Substituting this Ansatz into (2.32) and dividing by ψ we get [Har61]:

$$
\frac{1}{\rho R}\frac{\partial}{\partial \rho}\left(\rho\frac{\partial R}{\partial \rho} \right) + \frac{1}{\Phi}\frac{1}{\rho^2}\frac{\partial^2 \Phi}{\partial \varphi^2} + \frac{1}{Z}\frac{\partial^2 Z}{\partial z^2} + k^2 = 0
\tag{2.33}
$$

Using the constraint equation (2.34) and separation of variables, we get the following three differential equations, for each coordinate separately:

$$k_\rho^2 + k_z^2 \;=\; k^2 \tag{2.34}$$

$$\rho\frac{d}{d\rho}\left(\rho\frac{dR}{d\rho}\right) + [(k_\rho\rho)^2 - m^2]R \;=\; 0 \tag{2.35}$$

$$\frac{d^2\Phi}{d\varphi^2} + m^2\Phi \;=\; 0 \tag{2.36}$$

$$\frac{d^2Z}{dz^2} + k_z^2 Z \;=\; 0 \tag{2.37}$$

Equation (2.35) is known as Bessel's equation and the solutions to this equation are the different types of cylindrical functions, namely Bessel and Neumann functions[2]. Equations (2.36) and (2.37) are harmonic equations with cosine, sine and exponential functions as solutions. Starting with the result for R, we get:

$$R = A_1\, J_m(k_\rho\rho) + B_1\, N_m(k_\rho\rho) \tag{2.38}$$

and A_1 and B_1 being arbitrary constants. The Bessel function of order m can be defined as follows [AS64]

$$J_m(x) = \frac{j^{-m}}{\pi}\int_0^\pi e^{jx\cos\theta}\cos(m\theta)d\theta \tag{2.39}$$

while the Neumann function can be given as

$$N_m(x) = \frac{J_m(x)\cos(m\pi) - J_{-m}(x)}{\sin(m\pi)} \tag{2.40}$$

with

$$J_{-m}(x) = (-1)^m J_m(x) \quad \text{for integer m} \tag{2.41}$$

Since the Neumann function has a singularity at $\rho = 0$, the constant B_1 has to be zero, in order for the field to be finite for all ρ. The results for Φ and Z are:

$$\Phi \;=\; A_2 e^{jm\varphi} + B_2 e^{-jm\varphi} \tag{2.42}$$

$$Z \;=\; A_3 e^{jk_z z} + B_3 e^{-jk_z z} \tag{2.43}$$

---

[2]The use of Hankel functions as solutions to (2.35) is addressed in section 2.3.4

Since the field has to be 2π-periodic in φ, the order m has to be integer. Choosing the constants A_2, B_2, A_3, B_3 appropriately, we can construct traveling or standing waves in φ and z. Usually an azimuthal rotating wave is generated in the gyrotron cavity, we choose accordingly traveling waves in φ and z. This leads, with $A_1 = A_{mn}$, to the following solution for ψ:

$$\psi(\rho, \varphi, z) = A_{mn} J_m(k_\rho \rho) e^{\mp jm\varphi} e^{\mp jk_z z} \tag{2.44}$$

where the upper sign of the exponentials corresponds to clockwise rotating and in $+z$-direction traveling waves and the lower sign corresponds to counter-clockwise and in $-z$-direction traveling waves.
Applying the boundary conditions for TE modes results in letting

$$\frac{\partial J_m(k_\rho \rho)}{\partial(k_\rho \rho)} = 0 \bigg|_{\rho=R_0} \quad \Rightarrow \quad k_\rho = \frac{\chi'_{mn}}{R_0} \tag{2.45}$$

where χ'_{mn} is the root of equation (2.45) of order n.
For TM modes, applying the boundary condition gives:

$$J_m(k_\rho R_0) = 0 \Rightarrow k_\rho = \frac{\chi_{mn}}{R_0} \tag{2.46}$$

where χ_{mn} is the root of equation (2.46) of order n. Using (2.44) for F_z and substituting this into equations (2.20) and (2.21), we obtain the electric field components for TE modes:

$$
\begin{aligned}
E_\rho(\rho, \varphi, z) &= -\frac{1}{\varepsilon\rho}\frac{\partial\psi}{\partial\varphi} \\
&= \pm j\frac{m}{\varepsilon\rho} A_{mn} J_m(k_\rho\rho) e^{\mp jm\varphi} e^{\mp jk_z z} \tag{2.47} \\
E_\varphi(\rho, \varphi, z) &= \frac{1}{\varepsilon}\frac{\partial\psi}{\partial\rho} \\
&= \frac{k_\rho}{\varepsilon} A_{mn} \frac{\partial J_m(k_\rho\rho)}{\partial(k_\rho\rho)} e^{\mp jm\varphi} e^{\mp jk_z z} \tag{2.48} \\
E_z(\rho, \varphi, z) &= 0 \tag{2.49}
\end{aligned}
$$

The magnetic field components for TE modes are:

$$
\begin{aligned}
H_\rho(\rho,\varphi,z) &= -j\frac{1}{\omega\mu\varepsilon}\frac{\partial^2\psi}{\partial\rho\partial z} \\
&= \mp\omega\frac{k_\rho k_z}{k^2}A_{mn}\frac{\partial J_m(k_\rho\rho)}{\partial(k_\rho\rho)}e^{\mp jm\varphi}e^{\mp jk_z z} & (2.50)
\end{aligned}
$$

$$
\begin{aligned}
H_\varphi(\rho,\varphi,z) &= -j\frac{1}{\omega\mu\varepsilon\rho}\frac{\partial^2\psi}{\partial\varphi\partial z} \\
&= \mp j\omega\frac{m\cdot k_z}{k^2\rho}A_{mn}J_m(k_\rho\rho)e^{\mp jm\varphi}e^{\mp jk_z z} & (2.51)
\end{aligned}
$$

the sign depends on the product of the initial signs of m and k_z

$$
\begin{aligned}
H_z(\rho,\varphi,z) &= -j\frac{1}{\omega\mu\varepsilon}\left(\frac{\partial^2}{\partial z^2}+k^2\right)\psi \\
&= -j\frac{\omega k_\rho^2}{k^2}A_{mn}J_m(k_\rho\rho)e^{\mp jm\varphi}e^{\mp jk_z z} & (2.52)
\end{aligned}
$$

For TM modes, we substitute (2.44) for A_z in equations (2.22) and (2.23). In analogy to the constant A_{mn}, a constant B_{mn} is introduced in order to distinguish between TE and TM modes:

$$
\begin{aligned}
E_\rho(\rho,\varphi,z) &= -j\frac{1}{\omega\mu\varepsilon}\frac{\partial^2\psi}{\partial\rho\partial z} \\
&= \mp\omega\frac{k_\rho k_z}{k^2}B_{mn}\frac{\partial J_m(k_\rho\rho)}{\partial(k_\rho\rho)}e^{\mp jm\varphi}e^{\mp jk_z z} & (2.53)
\end{aligned}
$$

$$
\begin{aligned}
E_\varphi(\rho,\varphi,z) &= -j\frac{1}{\omega\mu\varepsilon\rho}\frac{\partial^2\psi}{\partial\varphi\partial z} \\
&= \mp j\omega\frac{m\cdot k_z}{k^2\rho}B_{mn}J_m(k_\rho\rho)e^{\mp jm\varphi}e^{\mp jk_z z} & (2.54)
\end{aligned}
$$

the sign depends on the product of the initial signs of m and k_z

$$
\begin{aligned}
E_z(\rho,\varphi,z) &= -j\frac{1}{\omega\mu\varepsilon}\left(\frac{\partial^2}{\partial z^2}+k^2\right)\psi \\
&= -j\frac{\omega k_\rho^2}{k^2}B_{mn}J_m(k_\rho\rho)e^{\mp jm\varphi}e^{\mp jk_z z} & (2.55)
\end{aligned}
$$

$$
\begin{aligned}
H_\rho(\rho, \varphi, z) &= -\frac{1}{\mu\rho}\frac{\partial\psi}{\partial\varphi} \\
&= \pm j\frac{m}{\mu\rho}B_{mn}J_m(k_\rho\rho)e^{\mp jm\varphi}e^{\mp jk_z z} \qquad (2.56)
\end{aligned}
$$

$$
\begin{aligned}
H_\varphi(\rho, \varphi, z) &= \frac{1}{\mu}\frac{\partial\psi}{\partial\rho} \\
&= \frac{k_\rho}{\mu}B_{mn}\frac{\partial J_m(k_\rho\rho)}{\partial(k_\rho\rho)}e^{\mp jm\varphi}e^{\mp jk_z z} \qquad (2.57)
\end{aligned}
$$

$$
H_z(\rho, \varphi, z) = 0 \qquad (2.58)
$$

We now have the solution for the field components in a circular cylindrical waveguide for one TE or TM mode rotating in $\pm\varphi$-direction and propagating in $\pm z$-direction.

Since the goal is to radiate the electromagnetic wave from the waveguide cut in form of a Gaussian beam (section 2.2), we need to obtain a mixture of different modes to form the Gaussian field profile. Consequently the electromagnetic wave will not consist of one single electromagnetic mode, but a linear combination of different modes. This can be achieved by summing over the azimuthal indices m and integrating over the longitudinal wave numbers k_z, see [Har61]:

$$
\psi(\rho, \varphi, z) = \sum_m \int_{k_z} A_{mn}(k_z)J_m(\sqrt{k^2 - k_z^2}\rho)e^{-jm\varphi}e^{-jk_z z}dk_z \qquad (2.59)
$$

The previous derivations were made for TE as well as TM modes. However in the following only TE modes will be considered, since the gyrotron interaction usually only excites TE modes effectively [Nus04]. The only source for TM modes is mode conversion, which will be neglected due to the smooth change in the waveguide surface.

2.1.3 Geometric-optical field representation in cylindrical waveguides

In order to get a better understanding of the wave propagation inside an overmoded waveguide, we can use a ray model (Fig. 2.2) that results from geometrical optics. We give a summary of the geometric optical parameters, a more detailed derivation can be found in [Dru02, Mic98, Wie95]. We can use the scalar potential from equation (2.44) to describe the wave propagation. By separating the Bessel function in (2.44) into the Hankel functions of first and second kind with order m, we represent the standing wave in radial direction with two traveling waves.

$$J_m(x) = \frac{1}{2}\left(H_m^{(1)}(x) + H_m^{(2)}(x)\right) \tag{2.60}$$

$$\psi = \psi^{(1)} + \psi^{(2)} = \frac{1}{2}\left[H_m^{(1)}(k_\rho\rho) + H_m^{(2)}(k_\rho\rho)\right] e^{\mp jm\varphi}e^{\mp jk_z z} \tag{2.61}$$

Furthermore, we take an adapted form of the approximation from [MF53a] for the two Hankel functions, that is valid for $x = \chi'_{mn} > m$ and large arguments x and hence valid on the waveguide wall:

$$H_m^{(1),(2)}(x) \approx \sqrt{\frac{2}{\pi\sqrt{x^2 - m^2}}} \cdot e^{\pm j\left(\sqrt{x^2-m^2}-m\cdot\arccos\left(\frac{m}{x}\right)-\frac{\pi}{4}\right)} \tag{2.62}$$

where the upper sign belongs to superscript (1) and the lower sign belongs to superscript (2). As we can see from (2.62):

$$H_m^{(1)}(x) = \left(H_m^{(2)}(x)\right)^* \tag{2.63}$$

If we now take the gradient of the phase function of the radially outgoing wave $\psi^{(2)}$ and use (2.62), we get an equation, that is similar to the Eikonal equation found in [BW09]. This gradient vector $\vec{R}$ is perpendicular to the phase front and represents the direction of the rays. For a wave propagating in positive z-direction and in positive φ-direction this results in:

$$\vec{R} = \nabla\left(\arg\left(\psi^{(2)}(R_0, \varphi, z)\right)\right) \tag{2.64}$$

$$= \nabla\left(-\sqrt{(k_\rho\rho)^2 - m^2} + m\arccos\left(\frac{m}{k_\rho\rho}\right) + \frac{\pi}{4} - m\varphi - k_z z\right)\Bigg|_{k_\rho\rho=\chi'_{mn}} \tag{2.65}$$

Considering figure 2.2 and projecting this ray $\vec{R}$ onto $\vec{e}_z$, we get the so-called Brillouin angle Ψ:

$$\cos\Psi = \frac{k_z}{k} \tag{2.66}$$

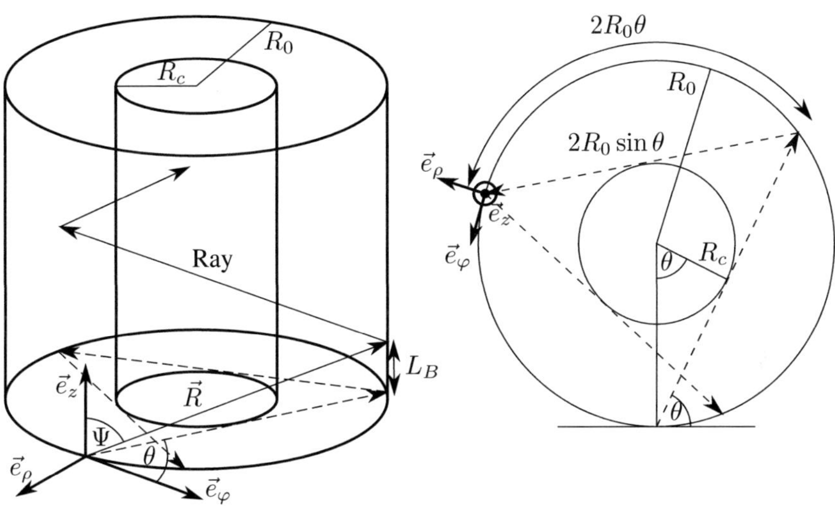

Figure 2.2: Ray modeling in a circular waveguide, left: side view perspective, right: top view

The wave number along the ray $\vec{R}$ is the free space wave number k. The distance the ray $\vec{R}$ travels in z-direction is called Brillouin length L_B shown in the left part of figure 2.2. This distance can be calculated from the TE mode's parameters as follows:

$$L_B = 2R_0 \sin\theta \cot\Psi \quad , \tag{2.67}$$

The projection in the transverse plane onto $\vec{e}_\varphi$ yields the spread angle θ as

$$\cos\theta = \frac{m}{\chi'_{mn}} = \frac{R_c}{R_0} \quad , \tag{2.68}$$

where R_c is the so called caustic radius [Wei69] and the corresponding distance traveled in transverse direction, L_{trans} is defined as

$$L_{trans} = 2R_0 \sin\theta. \tag{2.69}$$

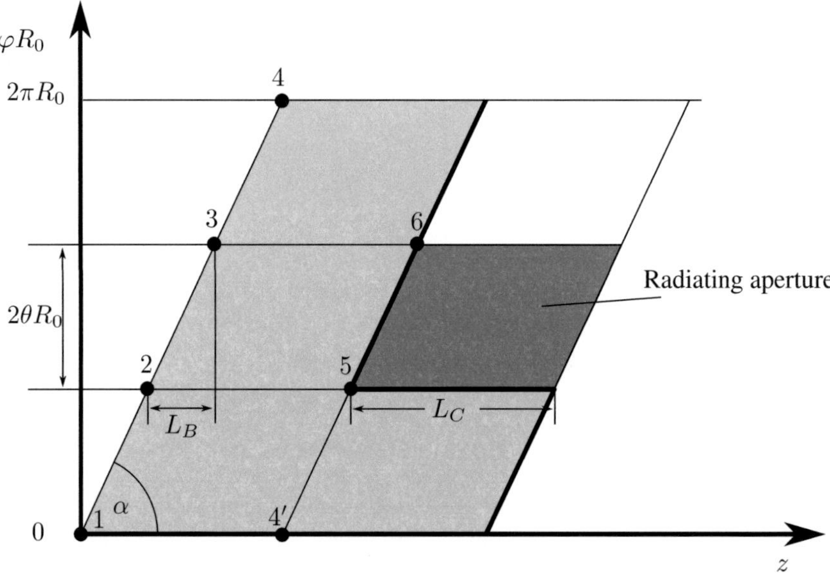

Figure 2.3: Reflections of the rays on the unrolled waveguide wall and Vlasov cut (thick line)

Along the circumference of the waveguide this length corresponds to:

$$L_\varphi = 2R_0\theta \tag{2.70}$$

The wave propagation can be described as follows: An outgoing ray represented by $\psi^{(2)}$ traveling towards the waveguide wall gets reflected from the metal surface and transforms into an ingoing ray represented by $\psi^{(1)}$. The ingoing ray experiences a phase jump, when reaching the caustic at R_c. This can be regarded as "transformation" of the ingoing ray into the outgoing ray, then the ray gets reflected again and so on. In this way the ray travels forward through the waveguide. To describe the whole field, we have to consider all rays traveling inside the waveguide. Figure 2.3 shows the unrolled waveguide surface at a constant

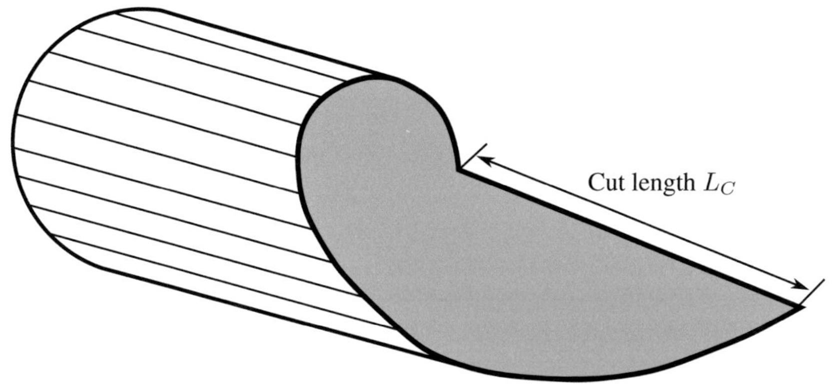

Figure 2.4: Launcher with helical cut of length L_C

radius R_0. Here we can see that all rays starting on the same helical line end on that line. So the ray starting at point 1 travels to point 2, then to point 3, then to point $4 \equiv 4'$ and so on. It can be shown [Wie95], that all rays touch the area contained within the parallelogram $\overrightarrow{2,3,5,6}$, with size $L_C \times 2\theta R_0$, once. This area is called Brillouin zone. If we cut the waveguide open at a constant angle of φ with length L_C, we radiate all rays as was stated before and therefore radiate all electromagnetic energy from the waveguide. As can be seen from [Wie95], the cut length of a waveguide antenna in order to radiate all rays can be calculated as

$$
\begin{aligned}
L_C &= \frac{2\pi R_0}{\tan \alpha} \\
&= 2\pi R_0 \frac{\sin \theta}{\theta} \cot \Psi
\end{aligned}
\tag{2.71}
$$

with

$$
\tan \alpha = \frac{2R_0 \theta}{L_B}
\tag{2.72}
$$

The waveguide antenna with this type of cut achieves the separation of the electron beam from the electromagnetic wave. Thus the design of the collector of the gyrotron can be made independent of the RF output system, which is important when going to megawatt power levels. This type of cut was first introduced by Vlasov [VZP75] and is commonly referred to as "Vlasov cut".

2.2 Field description for conventional launchers

The introduced form of a launcher (Fig. 2.4), a smooth waveguide with Vlasov cut, will radiate a constant aperture field and therefore has significant surface currents along the cut and edges of the waveguide. This leads to strong edge diffraction and consequently stray radiation inside the gyrotron.

In order to reduce this effect, the conventional method proposed in [DPV90] is to form a Gaussian-like field profile on the radiating aperture of the launcher by superimposing multiple waveguide modes. The modes necessary for the superposition are shown in table 2.1. A derivation of the necessary modes can be found in [Wie95] or [BD04]. To excite the necessary modes inside the waveguide,

$TE_{m-2,n+1}$ 3%	$TE_{m+1,n}$ 11%	$TE_{m+4,n-1}$ 3%
$TE_{m-3,n+1}$ 11%	$TE_{m,n}$ 44%	$TE_{m+3,n-1}$ 11%
$TE_{m-4,n+1}$ 3%	$TE_{m-1,n}$ 11%	$TE_{m+2,n-1}$ 3%

Table 2.1: TE modes and their relative powers involved in the mode mixture to form a Gaussian-like distribution

we start from the main mode generated in the cavity. We then modify the surface of the waveguide in order to excite these desired satellite modes. To describe the propagation of these multiple modes, we use the method of coupled modes, which is valid for small perturbations of the circular waveguide wall. For the case of small perturbations, reflections can be ignored and the equation for forward traveling modes is given as [Sol59, Thu84]:

$$\frac{\partial A_i(z)}{\partial z} = -jk_{z,i}A_i(z) + \sum_j K_{ij}(z)A_j(z) \tag{2.73}$$

A_i and A_j are the amplitudes of the modes at one z-position, the $k_{z,i}$ is the wave number in z-direction of the i-th mode and K_{ij} is the coupling coefficient from one mode to another. The solution to (2.73) gives the amplitudes of the different modes along the waveguide axis. The coupling coefficients K_{ij} result from a surface deformation of the following form

$$R(\varphi, z) = R_0 + \sum_{\Delta m} a_{\Delta m}(z) \cos(\Delta k_{z,\Delta m}z \pm \Delta m\varphi) \tag{2.74}$$

and have been derived in [Doa85, Sol59, BD04]. These coupling coefficients are non-zero for $|\Delta m| = |m_i - m_j|$ and $\Delta k_{z,\Delta m} = |k_{z,i} - k_{z,j}|$. So considering table 2.1, we need a deformation with $\Delta m = 1$ and $\Delta m = 3$, which directly excites the powerful modes (11 %), and as a secondary effect the weaker ones (3 %). We get for the wall deformation:

$$R(\varphi, z) = R_0 + \alpha z + a_1(z)\cos(\Delta k_{z,1}z - \varphi) + a_3(z)\cos(\Delta k_{z,3}z - 3\varphi) \tag{2.75}$$

Here a taper angle α is added to the radius to prevent spurious oscillation in the launcher [Thu97]. $\Delta k_{z,1}$ is the average difference of the wave numbers in z-direction of the $TE_{m,n}$ and $TE_{m\pm1,n}$ mode and $\Delta k_{z,3}$ is the average difference of the wave numbers in z-direction of the $TE_{m,n}$ and $TE_{m\pm3,n\mp1}$ mode. Since the superposition on the radiating aperture has to be in phase. The choice of modes has to result in a minimum in the initial phase difference. As a result, for the $\Delta m = 3$ modes, the radial index is changed by $\Delta n = \pm1$ in order to keep the difference $\Delta\chi'_{mn}$ between the main mode's Bessel root χ'_{mn} to the satellite modes', and therefore their phase difference to the main mode, at a minimum. This choice works efficiently for main modes with a spread angle $\theta \approx \frac{\pi}{3}$ as shown by the following explanation.

The difference in Bessel root can for high order modes ($m \gg 1$ and $n \gg 1$) be expressed by differentiation of Debye's approximation [BD04]:

$$\Delta\chi'_{mn} = \frac{\Delta m\theta + \Delta n\pi}{\sin\theta} \tag{2.76}$$

As we see from (2.76), the difference in Bessel root, compared to the main mode, depends on the spread angle and is therefore, see equation (2.68), dependent on the mode's azimuthal index and Bessel root of the the main mode. Choosing modes for the superposition with

$$\Delta m\theta \approx -\Delta n\pi \tag{2.77}$$

results in $\Delta\chi'_{mn}$ being very small and thus the phase difference between the necessary modes is small. Due to this result, $\theta \approx \frac{\pi}{3} = 60°$ is a typical choice for conventional gyrotron operating mode. As we will see later (Fig. 2.8), the chosen mode mixture is not equally good achievable in a reasonable launcher length for all gyrotron modes (due to the different spread angles of the operating mode).

2.2.1 Examples

Good field bunching

Figure 2.5 shows the result of the coupled mode calculation for a launcher with specifications given in table 2.2. The necessary amplitudes from equation 2.75 are shown in figure 2.6. This results in the H_z-component, plotted on the waveguide wall over φ and z. The commercial program LOT [Nei04] was used to calculate this result. Figure 2.2 illustrates the three reflections along the circumference of the waveguide for the $TE_{22,6}$ mode with a spread angle of $\theta = 61.17°$. All modes with approximately this spread angle show a good conversion into a Gaussian field profile shown in table 2.1. The aforementioned difference in Bessel root for this example is ($\Delta m = \pm 3, \Delta n = \mp 1$):

$$\Delta\chi'_{22,6} = \frac{\pm 3 \cdot \theta \mp \pi}{\sin\theta} \approx \pm 0.06998 \tag{2.78}$$

From figure 2.5, we can see that the field starts with a homogeneous azimuthal distribution at the beginning of the launcher and then increases the bunching of the field profile, as we go towards the end of the launcher. The white line in the figure indicates the cut of the launcher, the focused field directly after the cut is radiated obliquely into free space and onto the following mirrors. In order to determine the conversion efficiency to the Gaussian field shape, we correlate the radiated field to

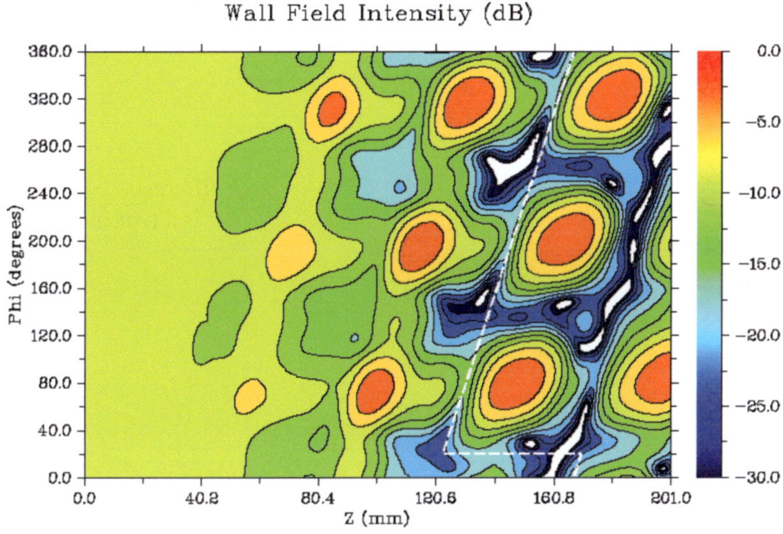

Figure 2.5: H_z-component on the waveguide wall for the $TE_{22,6}$ mode

Parameter	Value
Operating mode	$TE_{22,6}$
Frequency f	118 GHz
Brillouin angle Ψ	67.28 °
Spread angle θ	61.17 °
Launcher radius R_0	20 mm
Taper angle $\tan \alpha$	0.002

Table 2.2: Parameters of the launcher used for the result in figure 2.5

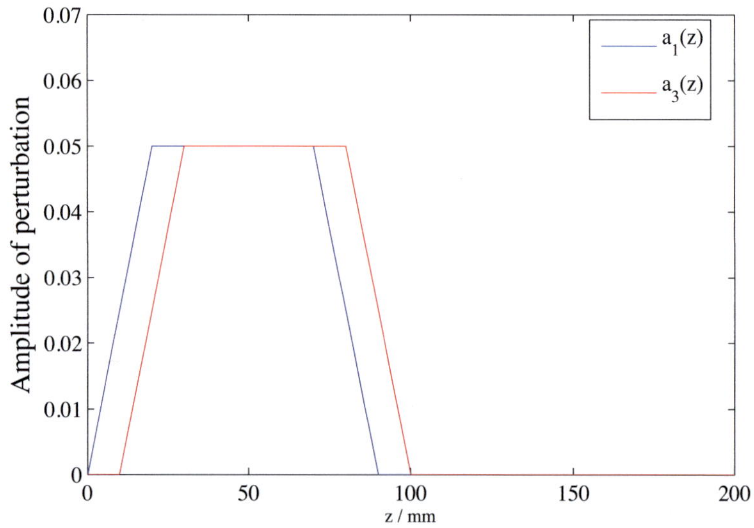

Figure 2.6: Amplitudes of the wall deformation for the $TE_{22,6}$-mode launcher

a Gaussian field profile. This can be done using the following two metrics:

$$\eta_{scalar} = \frac{\left(\iint\limits_{S} |u_1| \cdot |u_2| \, dS \right)^2}{\iint\limits_{S} |u_1|^2 \, dS \cdot \iint\limits_{S} |u_2|^2 \, dS} \tag{2.79}$$

$$\eta_{vector} = \frac{\left| \iint\limits_{S} u_1^* \cdot u_2 \, dS \right|^2}{\iint\limits_{S} |u_1|^2 \, dS \cdot \iint\limits_{S} |u_2|^2 \, dS} \tag{2.80}$$

where S is the aperture surface.
Equation (2.79) compares the amplitudes of two field components u_1 and u_2

with respect to power, whereas equation (2.80) includes the phase of the field components in the comparison. For this example the scalar Gaussian correlation is $\eta_{scalar} = 96.77\%$ and the complex Gaussian correlation is $\eta_{vector} = 94\%$ [PADT07]. In [Pri07] a version of this launcher was further optimized. The complex Gaussian correlation of the optimized version is augmented to $\eta_{vector} = 95\%$ and the field amplitudes on the cut are further reduced by 7 dB.

Insufficient field bunching

Now if we consider a launcher with an operating mode for a coaxial cavity gyrotron, with parameters shown in table 2.3 and perturbation amplitudes given in figure 2.7, we notice the different spread angle of $\theta = 71.14\,°$. This spread angle results in 2.5 consecutive reflections on the waveguide wall to achieve a full turn. In order to generate the satellite modes, needed according to table 2.1 and taking equation (2.77) into account, we get for $\Delta n = -1$ and $\Delta m = 3$:

$$\Delta\chi'_{34,19} = \frac{3 \cdot \theta - \pi}{\sin \theta} \approx 0.616 \neq 0 \tag{2.81}$$

For the $TE_{22,6}$ mode the difference in Bessel roots $\Delta\chi'_{mn}$ is smaller by nearly a factor of 10. The resulting phase difference of the satellite mode is not negligible compared to the case of the $TE_{22,6}$ mode ($\theta = 61.17\,°$). This results in insufficient bunching even after a launcher length of 450 mm (Figure 2.8) and a small launcher radius of 31.03 mm. The white line as before shows the cut of the launcher. The scalar correlation to the Gaussian field profile for this launcher results in $\eta_{scalar} = 86.51\,\%$. For the launcher and the mirror system, a complex vector correlation of $\eta_{vector} = 90\,\%$ was determined in [Rze07]. For high power gyrotrons, this launcher design is unacceptable due to the resulting stray radiation. One possibility to solve this problem is to use an adaptive wall deformation in order to convert the field over a shorter distance and with better efficiency. An approach, using higher harmonics for the surface perturbation, to model the adaptive wall is very impractical, when employing the coupled mode equations due to the amount of necessary harmonic surface perturbation to describe the wall of the waveguide.

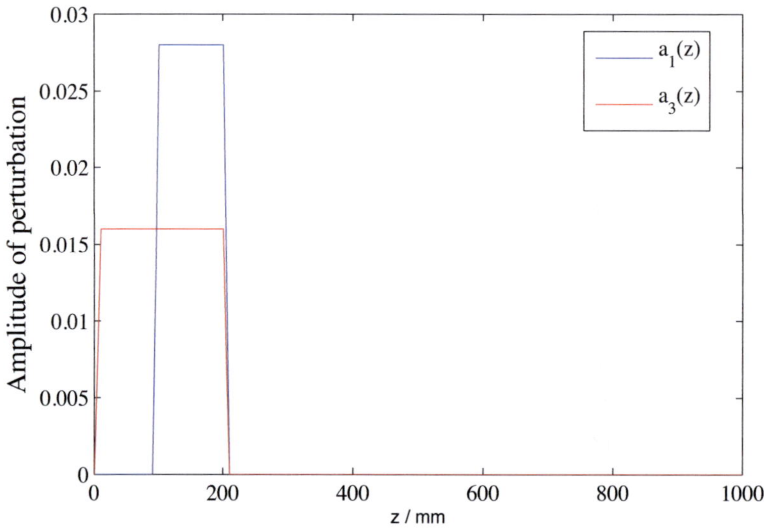

Figure 2.7: Amplitudes of the wall deformation for the $TE_{34,19}$-mode launcher

Parameter	Value
Operating mode	$TE_{34,19}$
Frequency f	170 GHz
Brillouin angle Ψ	65.29 °
Spread angle θ	71.14 °
Launcher radius R_0	31.03 mm
Taper angle $\tan \alpha$	0.002

Table 2.3: Parameters of the launcher used for the result in figure 2.8

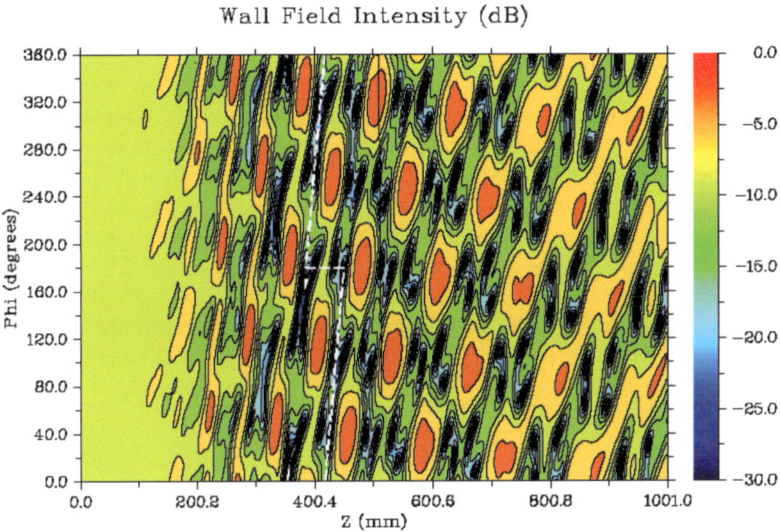

Figure 2.8: H_z on the waveguide wall for the $TE_{34,19}$ mode

2.3 Diffraction in waveguide antennas

One way to overcome the shortcomings of the coupled mode method in the
last section and use a better suited wall deformation is to solve the Helmholtz
equation (2.31) by separating the standing wave in radial direction into two radially
traveling waves and model the waveguide wall as a reflector. The perturbation of
the waveguide wall can then (with certain restrictions) be interpreted as a phase
corrector. This is similar to the geometric-optical representation from section
2.1.3. The difference to the former representation is the additional consideration
of fast phase correction and diffraction occurring inside the waveguide. This
approach corresponds to physical optics and is called scalar quasi-optical method.
We will now derive the necessary formulas.

31

We start from equation (2.31) and apply the decomposition of the potential:

$$\psi = \psi^{(2)} + \psi^{(1)} \tag{2.82}$$

where $\psi^{(2)}$ being the part traveling radially outwards and $\psi^{(1)}$ radially inwards. We start from $\psi^{(1)} \equiv u$ as the potential on the wall propagating away from the wall towards the inside of the waveguide. Due to linearity of the partial differential equation (2.31), u also fulfills the Helmholtz equation. So we have:

$$\nabla^2 u + k^2 u = 0 \tag{2.83}$$

Now we use the Green's function formalism to reformulate this equation. For more details on Green's functions see [MF53a, MF53b, Har61]. The Green's function fulfills the same differential equation except for the singularity at $\vec{r} - \vec{r}' = 0$:

$$\nabla^2 G + k^2 G = -\delta\left(\vec{r} - \vec{r}'\right) \tag{2.84}$$

where δ denotes the Dirac delta function [Goo05]. The form of the Green's function will be considered later. Now multiplying (2.83) with G and (2.84) with u, subtracting one equation from the other and integrating over the whole volume V (inside the waveguide), we get:

$$\iiint_V u\left(\vec{r}\right) \delta\left(\vec{r} - \vec{r}'\right) dV = $$
$$\iiint_V \left(G\left(\vec{r}, \vec{r}'\right) \nabla^2 u\left(\vec{r}\right)\right) - u\left(\vec{r}\right) \nabla^2 G\left(\vec{r}, \vec{r}'\right)\right) dV \tag{2.85}$$

Application of Green's theorem

$$\iiint_V \left[\xi \nabla^2 \zeta - \zeta \nabla^2 \xi\right] dV = \iint_S \left(\xi \nabla \zeta - \zeta \nabla \xi\right) d\vec{S}$$

to the right hand side of (2.85) and integrating the left hand side

$$u(\vec{r}') = \iiint_V u\left(\vec{r}\right) \delta\left(\vec{r} - \vec{r}'\right) dV$$

yields, after interchanging the coordinates[3], the scalar diffraction integral known as:

$$u(\vec{r}) = - \iint\limits_{S} \left[u\left(\vec{r}\,'\right) \nabla\,'G\left(\vec{r},\,\vec{r}\,'\right) - G\left(\vec{r},\,\vec{r}\,'\right) \nabla\,'u\left(\vec{r}\,'\right) \right] \vec{n}\,' \, dS' \qquad (2.86)$$

Here $\nabla\,'$ means the differentiation applies only to primed (=source) coordinates not observation coordinates (unprimed coordinates) and $\vec{n}\,'$ denotes the normal vector to the surface S (Fig. 2.9). This integral allows us to determine the field component $u \equiv \psi^{(1)}$ inside the volume from the values of $u(\vec{r}\,')$ and $\nabla\,'u(\vec{r}\,')$ over the waveguide surface S enclosing the volume V (Fig. 2.9). The arrows in figure 2.9 indicate the direction of propagation of $u(\vec{r}\,')$. The resulting $u(\vec{r})$ will be equivalent to $\psi^{(2)}$. This naming convention is valid for the Dipole Source Diffraction (DSD) in subsection 2.3.2 and for the Cylindrical Wave Decomposition (CWD) explained in subsection 2.3.4. For the following Point Source Diffraction (PSD) the potential $u \equiv \psi$ is used as introduced in [JTP+09]. The surface used for the integration will be the smooth waveguide surface. The perturbation is implemented as a phase corrector (see subsection 2.3.3). The condition for this scalar treatment of the diffraction inside oversized waveguides, according to (2.86), to be valid is that the electromagnetic wave remains paraxial [CDK+06], resulting in the condition $kR_0 \gg 1$. Further restrictions of the scalar theory will be discussed when necessary.

2.3.1 Point Source Diffraction (PSD)

In this section diffraction using a point source is summarized, for details see [JTP+09]. The boundary condition for the potential $\psi \equiv u$ is:

$$\nabla\,'\psi \cdot \vec{n}\,' = \left. \frac{\partial \psi}{\partial n'} \right|_{\rho = R_0} = 0 \qquad (2.87)$$

One solution to equation (2.86) can be found when choosing a point source as the Green's function also known as the free space Green's function [Bal89].

$$G(\vec{r}, \vec{r}\,') = G_K(\vec{r}, \vec{r}\,') = \frac{1}{4\pi} \frac{e^{-jkr_0}}{r_0} \quad , r_0 = |\vec{r}_0| = |\vec{r} - \vec{r}\,'| \qquad (2.88)$$

---

[3]possible due to symmetry of G

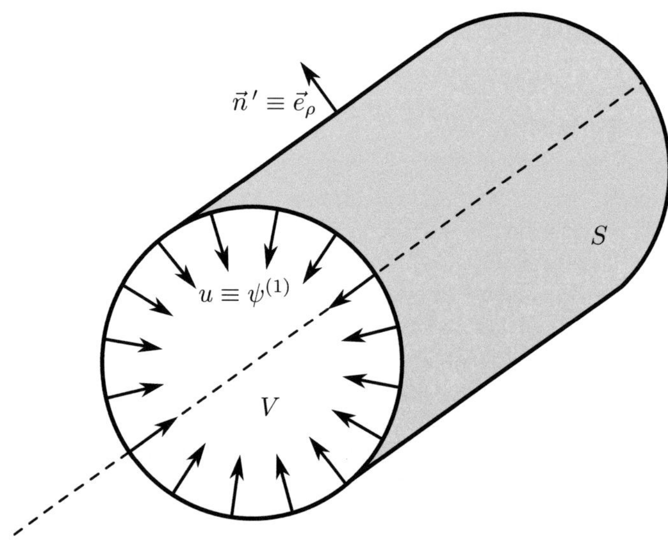

Figure 2.9: Geometry for the integration contours for the scalar diffraction integral

This function fulfills equation (2.84). With the given boundary condition for ψ, equation (2.86) gives:

$$u(\vec{r}) = - \iint\limits_{S} u\left(\vec{r}\,'\right) \nabla'G\left(\vec{r},\,\vec{r}\,'\right) \vec{n}\,' \, dS' \tag{2.89}$$

For the derivative of the Green's function we obtain ([Bro05]):

$$\begin{aligned}
\nabla'G \cdot \vec{n}\,' &= \frac{1}{4\pi} \frac{d}{dr_0}\left(\frac{e^{-jkr_0}}{r_0}\right) \vec{n}\,' \cdot \frac{\vec{r}_0}{r_0} \\
&= -\frac{1}{4\pi}\left(jk + \frac{1}{r_0}\right) \cdot \frac{e^{-jkr_0}}{r_0} \vec{n}\,' \cdot \frac{\vec{r}_0}{r_0} \tag{2.90}
\end{aligned}$$

In a cylindrical coordinate system this gives at constant radius $\rho = R_0$ (see appendix A.1):

$$r_0 \;=\; \sqrt{(z - z')^2 + 4R_0^2 \sin^2\left(\frac{\varphi - \varphi'}{2}\right)} \tag{2.91}$$

$$\nabla' G \cdot \vec{n}' \;=\; \frac{1}{4\pi}\left(jk + \frac{1}{r_0}\right) \cdot \frac{-2 \cdot R_0 \cdot \sin^2\left(\frac{\varphi - \varphi'}{2}\right)}{r_0} \cdot \frac{e^{-jkr_0}}{r_0} \tag{2.92}$$

$$dS' \;=\; R_0\, d\varphi'\, dz' \tag{2.93}$$

Inserting (2.92) into equation (2.89) gives:

$$u(\vec{r}) \;=\; \frac{1}{4\pi} \int\limits_{-\infty}^{L} \int\limits_{0}^{2\pi} u(\vec{r}') \cdot \left(\frac{e^{-jkr_0}}{r_0^2}\right)\left(jk + \frac{1}{r_0}\right) \cdot$$
$$2 \cdot R_0 \cdot \sin^2\left(\frac{\varphi - \varphi'}{2}\right) R_0 d\varphi'\, dz' \tag{2.94}$$

This equation is equation (11) in [JTP$^+$09] and gives the solution of a launcher of length L. Equation (2.94) lacks the phase corrector this will be introduced later as mentioned before.

2.3.2 Dipole Source Diffraction (DSD)

Choosing a different Green's function in order to consider only the inward ($\psi^{(1)}$) traveling part of ψ and overcoming the problem of needing the potential and its derivative, since

$$\frac{\partial u}{\partial n'} \equiv \frac{\partial \psi^{(1)}}{\partial n'} \neq 0, \tag{2.95}$$

we follow the derivation and choose according to [Som89]:

$$G_{RS}(\vec{r}_{1*}, \vec{r}_1, \vec{r}') = \frac{1}{4\pi}\left(\frac{e^{-jk|\vec{r}_1 - \vec{r}'|}}{|\vec{r}_1 - \vec{r}'|} - \frac{e^{-jk|\vec{r}_{1*} - \vec{r}'|}}{|\vec{r}_{1*} - \vec{r}'|}\right) \tag{2.96}$$

This corresponds to two point sources with a $180\,°$ phase shift. For the two sources we choose the vector $\vec{r}'$ to be located symmetrically along the surface S between

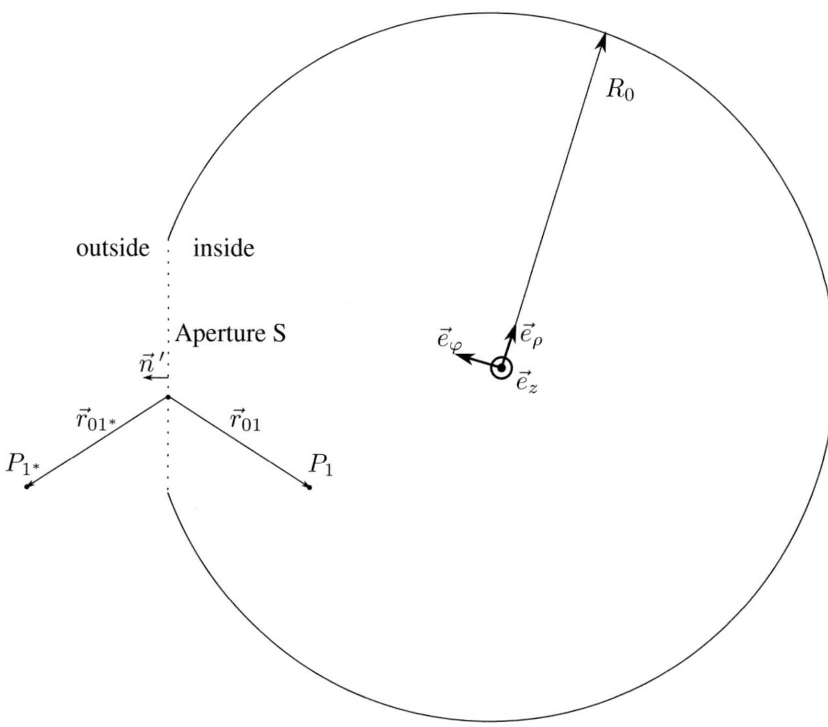

Figure 2.10: Construction of the Green's function for Rayleigh-Sommerfeld diffraction at an aperture

the two points P_1 and P_{1*} as shown in figure 2.10. This results in two vectors, $\vec{r}_{01} = \vec{r}_0 = \vec{r}_1 - \vec{r}'$ and $\vec{r}_{01*} = \vec{r}_{1*} - \vec{r}'$. The right part of figure 2.10 represents the inside, the left part represents the outside of the oversized waveguide, when calculating the field in the waveguide using this type of Green's function. As can be seen from figure 2.10:

$$|\vec{r}_{01}| = r_{01} = |\vec{r}_{01*}| = r_{01*} = r_0 \qquad\qquad , \vec{r}' \text{on } S \qquad (2.97)$$

and therefore:

$$G_{RS} = 0 \quad \text{on } S \tag{2.98}$$

With this result, equation (2.86) reduces to

$$u(\vec{r}) = - \iint\limits_{S} u(\vec{r}') \, \nabla' G_{RS}(\vec{r}, \vec{r}') \, \vec{n}' \, dS' \tag{2.99}$$

which is identical to (2.89) in form, but different since G_{RS} and its derivative are different.

The derivative of G_{RS} can be obtained through:

$$\begin{aligned}
\nabla' G_{RS} \cdot \vec{n}' &= \frac{\partial G_{RS}}{\partial n'} \\
&= \frac{d}{dr_{01}} \left(\frac{e^{-jkr_{01}}}{r_{01}} \right) \frac{\partial r_{01}}{\partial n'} \\
&\quad - \frac{d}{dr_{01*}} \left(\frac{e^{-jkr_{01*}}}{r_{01*}} \right) \frac{\partial r_{01*}}{\partial n'},
\end{aligned} \tag{2.100}$$

$$r_0 = r_{01} = r_{01*}, \tag{2.101}$$

$$\frac{\partial r_{01}}{\partial n'} = -\frac{\partial r_{01*}}{\partial n'} = \vec{n}' \cdot \frac{\vec{r}_0}{r_0}, \tag{2.102}$$

resulting finally in:

$$\nabla' G_{RS} \cdot \vec{n}' = -2 \cdot \frac{1}{4\pi} \left(jk + \frac{1}{r_0} \right) \cdot \frac{e^{-jkr_0}}{r_0} \vec{n}' \cdot \frac{\vec{r}_0}{r_0} \tag{2.103}$$

Comparing (2.103) with (2.90) gives the difference between the two types of diffraction:

$$\nabla' G_{RS} \cdot \vec{n}' = 2 \cdot \nabla' G_K \cdot \vec{n}' \tag{2.104}$$

Inserting this into (2.99) and writing (2.103) in cylindrical coordinates gives the result for the cylindrical waveguide:

$$\begin{aligned}
u(\vec{r}) &= \frac{1}{4\pi} \int\limits_{0}^{2\pi} \int\limits_{-\infty}^{L(\varphi)} u(\vec{r}') \cdot \left(\frac{e^{-jkr_0}}{r_0^2} \right) \left(jk + \frac{1}{r_0} \right) \cdot \\
&\quad 4 \cdot R_0 \cdot \sin^2 \left(\frac{\varphi - \varphi'}{2} \right) R_0 dz' d\varphi'
\end{aligned} \tag{2.105}$$

with r_0 as given in equation (2.91) and

$$\nabla' G_{RS} \cdot \vec{n}' = \frac{1}{4\pi}\left(jk + \frac{1}{r_0}\right) \cdot \frac{-4 \cdot R_0 \cdot \sin^2\left(\frac{\varphi - \varphi'}{2}\right)}{r_0} \cdot \frac{e^{-jkr_0}}{r_0}$$

$$dS' = R_0\, d\varphi'\, dz'$$

Due to the helical cut of the launcher with length L, the upper integration limit for z' is $L(\varphi)$ and depends on the azimuthal angle φ. Equation (2.105) is similar to the one given in [CDK$^+$06].

As known from scalar diffraction theory the potential u has to satisfy the Sommerfeld radiation condition [Som92]:

$$\lim_{r_0 \to \infty} r_0\left(\frac{\partial u}{\partial r_0} + jku\right) = 0 \quad \text{or} \quad \lim_{r_0 \to \infty} r_0\left(\frac{\partial u}{\partial r_0} - jku\right) = 0 \qquad (2.106)$$

The type of condition depends on the type of wave taken into account (either inward or outward traveling waves). Since we separate ψ into two different parts, the inward and the outward traveling part, our choice of u satisfies the radiation condition. This is not obvious[4] for $u \equiv \psi$ in equation (2.89). But within the framework of the scalar approach, solutions to equation (2.89) give reasonable results and have been verified with the commercial code Surf3D [Nei04]. For a derivation of the scalar integral equation with $u \equiv \psi^{(1)}$ using a point source and satisfying the radiation condition, see [Sil86]. As in section 2.3.1, equation (2.105) does not contain the perturbation of the waveguide wall. The described method corresponds to propagation of the electromagnetic wave along the wall of the un-perturbed oversized waveguide. The usage of (2.105) to calculate the field in a oversized circular waveguide has some restrictions:

- As mentioned before the scalar treatment of diffraction requires $kR_0 \gg 1$

- For the Green's function construction to be valid, the plane wave approximation for the Hankel function (2.62) has to hold, which requires $k_\rho R_0 > m$. For whispering gallery modes, where the field is concentrated near the waveguide wall, this condition is not necessarily fulfilled.

---

[4]For a discussion see [BC53] p.26-28

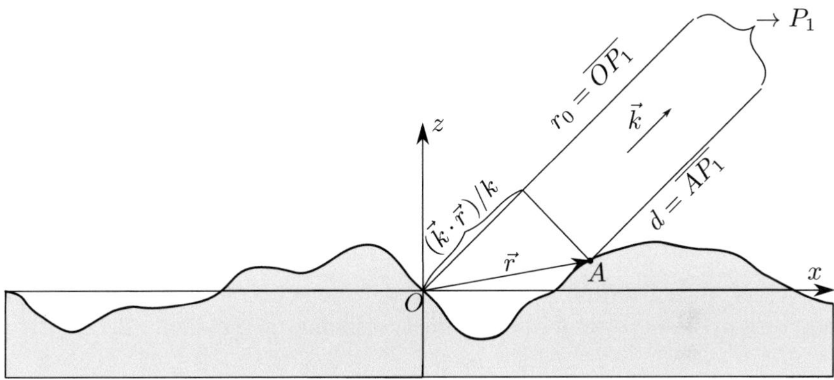

Figure 2.11: Geometry of a surface with perturbations

- The scalar treatment also fails, if the surface perturbation (explained in the next subsection) is not smooth enough. A discussion on the range of validity due to smoothness of the surface can be found in the appendix of [BS63].

2.3.3 Phase corrector model for small perturbations

Figure 2.11 shows a small surface perturbation with the observation point P_1 at a large distance. In this case, we can separate the distance d (being the actual calculation distance) into two parts. The first part is the propagation through the waveguide and the second part is the change in the field due to the perturbation[5]. Since the perturbation is supposed to be small, its effect on the amplitude is neglected. From figure 2.11 one can identify the first part as the distance r_0, used in the integral equations (2.94) and (2.105), and $(\vec{k}\cdot\vec{r})/k$ as the difference projected onto r_0 compared to $d = \overline{AP_1}$. Applying this to the exponential term

---

[5] Applying the change in field due to the perturbation after the propagation will result in a change of signs.

of the Green's function gives:

$$k(r_0 - d) = \vec{k} \cdot \vec{r}$$

$$\frac{e^{-jkd}}{d} \approx \frac{e^{-jkr_0 + j\vec{k}\cdot\vec{r}}}{r_0} = \underbrace{\frac{e^{-jkr_0}}{r_0}}_{\text{propagation}} \cdot \overbrace{e^{j\vec{k}\cdot\vec{r}}}^{\text{phase corrector}} \qquad (2.107)$$

In the case of an oversized waveguide with a radius perturbation $\Delta R(\varphi, z)$, $\vec{r} \equiv \vec{n}$ and with $\vec{k}$ pointing towards the perturbation surface, we get (figure 2.12(a)):

$$\vartheta(\varphi, z) = \vec{k} \cdot \vec{r} = \vec{k} \cdot \vec{n} = 2\left|\vec{k}\right| \Delta R(\varphi, z) \cos \alpha \qquad (2.108)$$

$$e^{j\vec{k}\cdot\vec{r}} = e^{-j2\left|\vec{k}\right|\Delta R \cos \alpha} \qquad (2.109)$$

$\cos \alpha$ denotes the directional cosine. The $\left|\vec{k}\right| \cos \alpha$ inside the waveguide can be expressed by the waveguide mode's parameters as follows:

- The projection of the wave vector $\vec{k}$ onto the (ρ, z)-plane gives (figure 2.12(b)):

$$\cos \alpha = \sin \theta = \sqrt{1 - \left(\frac{m}{\chi'_{mn}}\right)^2}$$

- The projection onto the (ρ, φ)-plane gives $\left|\vec{k}\right| = k_\rho$.

$$\Rightarrow \left|\vec{k}\right| \cos \alpha = k_\rho \sqrt{1 - \left(\frac{m}{\chi'_{mn}}\right)^2}$$

So the phase corrector on the waveguide wall can be expressed as:

$$e^{-j2\left|\vec{k}\right|\Delta R \cos \alpha} = e^{-j2k_\rho \Delta R \sqrt{1 - \left(\frac{m}{\chi'_{mn}}\right)^2}} \qquad (2.110)$$

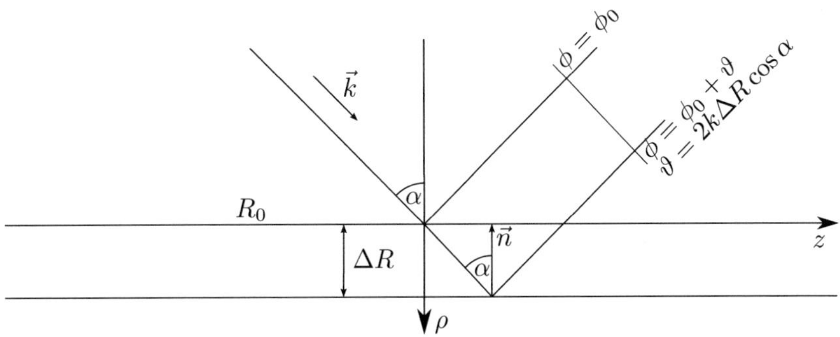

(a) Phase correction due to surface perturbation ΔR

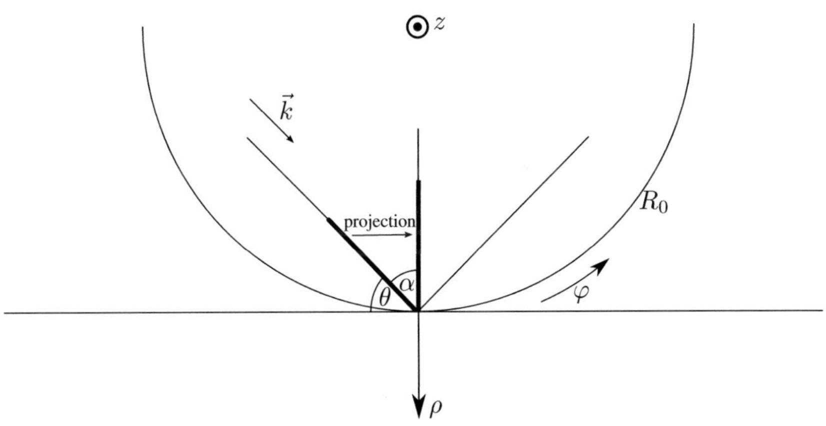

(b) Illustration of the projection of the wave vector onto the (ρ, z)-plane

Figure 2.12: Phase correction at the waveguide wall

Equations (2.94) and (2.105) give, incorporating the phase corrector[6]:

$$
u^{(P)}(R_0, \varphi, z) = e^{-j2k_\rho \Delta R(\varphi,z)\sqrt{1-\left(\frac{m}{\chi'_{mn}}\right)^2}} \cdot \frac{1}{4\pi} \int\limits_0^{2\pi} \int\limits_{-\infty}^{L(\varphi)} u^{(P)}(R_0, \varphi', z,') \cdot
$$

$$
\left(\frac{e^{-jkr_0}}{r_0^2}\right)\left(jk + \frac{1}{r_0}\right) \cdot
$$

$$
2 \cdot R_0 \cdot \sin^2\left(\frac{\varphi - \varphi'}{2}\right) R_0 dz' d\varphi' \tag{2.111}
$$

$$
u^{(D)}(R_0, \varphi, z) = e^{-j2k_\rho \Delta R(\varphi,z)\sqrt{1-\left(\frac{m}{\chi'_{mn}}\right)^2}} \cdot \frac{1}{4\pi} \int\limits_0^{2\pi} \int\limits_{-\infty}^{L(\varphi)} u^{(D)}(R_0, \varphi', z,') \cdot
$$

$$
\left(\frac{e^{-jkr_0}}{r_0^2}\right)\left(jk + \frac{1}{r_0}\right) \cdot
$$

$$
4 \cdot R_0 \cdot \sin^2\left(\frac{\varphi - \varphi'}{2}\right) R_0 dz' d\varphi' \tag{2.112}
$$

In equations (2.94) and (2.105) the integration can be regarded as an operator on the given domain. This gives:

$$
u^{(P)} = \mathcal{L}_1 u^{(P)} \tag{2.113}
$$

$$
u^{(D)} = \mathcal{L}_2 u^{(D)} \tag{2.114}
$$

The operator can usually not be obtained in a closed form, this is only possible in rare cases (e.g. Fresnel diffraction leads to the closed form known as Gaussian beams). So in order to calculate the field in the perturbed oversized waveguide, we solve the operator equation iteratively as described in subsection 2.3.5.

2.3.4 Cylindrical Wave Decomposition (CWD)

As the third method for the comparison of different methods to solve the diffraction problem in an oversized waveguide with perturbation, we use a superposition

---

[6]Superscript P denotes the PSD solution, Superscript D denotes the DSD solution

of multiple elementary wave functions [Har61], that each solves the Helmholtz equation. We follow our explanation of the method as detailed in [FJT11]. One example for a cylindrical geometry of such a superposition is as follows:

$$\psi = \sum_m \int_{k_z} g_m(k_z) B_m(k_\rho \rho) h(m\varphi) h(k_z z) dk_z \tag{2.115}$$

In this case $B_m(k_\rho \rho)$ are solutions to Bessel's equation and the functions $h(k_z z)$ and $h(m\varphi)$ are exponentials as introduced in subsection 2.1.2. As was shown before in (2.60), we separate the Bessel function $J_m(x)$ into the two Hankel functions. This enables us as before to treat the waveguide wall as a reflector and leads us to two cylindrical waves, one represented by the Hankel function of first kind $H_m^{(1)}(\rho\sqrt{k^2 - k_z^2})$ going inward and by the Hankel function of second kind $H_m^{(2)}(\rho\sqrt{k^2 - k_z^2})$ going outward, inside the waveguide.

$$\psi = \psi^{(2)} + \psi^{(1)} \tag{2.116}$$

with

$$\psi^{(2)}(\rho, \varphi, z) = \frac{1}{4\pi^2} \sum_{m=-\infty}^{\infty} e^{jm\varphi} \int_{-\infty}^{\infty} g_m^{(2)}(k_z) \cdot$$
$$H_m^{(2)}\left(\rho\sqrt{k^2 - k_z^2}\right) e^{jk_z z} dk_z \tag{2.117}$$

$$\psi^{(1)}(\rho, \varphi, z) = \frac{1}{4\pi^2} \sum_{m=-\infty}^{\infty} e^{jm\varphi} \int_{-\infty}^{\infty} g_m^{(1)}(k_z) \cdot$$
$$H_m^{(1)}\left(\rho\sqrt{k^2 - k_z^2}\right) e^{jk_z z} dk_z \tag{2.118}$$

where $\psi^{(2)}$ represents the total outward traveling wave and $\psi^{(1)}$ the total inward traveling wave. Equations (2.117) and (2.118) can easily be identified as Fourier sums and integrals. The wave amplitudes can be determined by performing the inverse operation at a constant initial radius $\rho = R_0$:

$$g_m^{(1),(2)}(k_z) = \frac{1}{H_m^{(1),(2)}\left(R_0\sqrt{k^2 - k_z^2}\right)} \cdot$$

$$\int\limits_0^{2\pi}\int\limits_{-\infty}^{\infty} \psi^{(1),(2)}(R_0, \varphi, z) \cdot e^{-jk_z z} e^{-jm\varphi} \, dz \, d\varphi \qquad (2.119)$$

Due to a singularity of the Hankel functions $H_m^{(1),(2)}$ for $\rho = 0$, the wave amplitudes have to be equal for the field to be finite everywhere: $g_m^{(1)}(k_z) = g_m^{(2)}(k_z)$. With this condition the outward traveling wave at the new position can be calculated from the inward traveling wave at the original position and we get:

$$\psi^{(2)}(\rho, \varphi, z) = \frac{1}{4\pi^2} \sum_{m=-\infty}^{\infty} \int\limits_{-\infty}^{\infty} \frac{H_m^{(2)}\left(\rho\sqrt{k^2 - k_z^2}\right)}{H_m^{(1)}\left(R_0\sqrt{k^2 - k_z^2}\right)} \cdot$$

$$\int\limits_0^{2\pi}\int\limits_{-\infty}^{\infty} \psi^{(1)}(R_0, \varphi', z') \cdot e^{-jk_z z'} e^{-jm\varphi'} \, dz' \, d\varphi' \, e^{jm\varphi} e^{jk_z z} \, dk_z \qquad (2.120)$$

Using equation (2.120) we can calculate the outgoing cylindrical wave field from an initial ingoing wave field. If we choose the outgoing radius to be $\rho = R_0$, we can describe the propagation along the straight smooth waveguide wall. We use a phase corrector in order to incorporate the slight mode converting perturbations ΔR of the waveguide wall and with the corresponding incident angle α of the wave with the surface, we get the resulting phase shift as shown in (2.122). By using the phase corrector model, the propagation is only applicable if the mode spectrum stays in the proximity of the initial $k_\rho = \sqrt{k^2 - k_z^2}$. That is the perturbations have to be smooth.

$$\psi^{(1)}(\rho, \varphi, z) = e^{-j\vartheta(\varphi, z)} \cdot \psi^{(2)}(\rho, \varphi, z) \qquad (2.121)$$

$$\vartheta(\varphi, z) = 2k_\rho \Delta R(\varphi, z)\sqrt{1 - \left(\frac{m}{\chi'_{mn}}\right)^2} \qquad (2.122)$$

This results in an operator equation with respect to the unknown $\psi^{(1)}$:

$$\psi^{(1)} = \mathcal{L}_3 \psi^{(1)} \tag{2.123}$$

$$\mathcal{L}_3 \psi^{(1)} = e^{-j\vartheta(\varphi,z)} \cdot \mathcal{F}^{-1} \mathcal{T} \cdot \mathcal{F} \psi^{(1)} \tag{2.124}$$

$$\mathcal{T} = \frac{H_m^{(2)}\left(R_0\sqrt{k^2 - k_z^2}\right)}{H_m^{(1)}\left(R_0\sqrt{k^2 - k_z^2}\right)} \tag{2.125}$$

where $\mathcal{F}$ represents the Fourier Transform and $\mathcal{F}^{-1}$ its inverse. This derivation is similar to [Kuz97]. In [PV07] the propagation from one cylindrical mirror to another is also described by cylindrical waves (Hankel functions).

In contrast to [Kuz97] the field structure does not have to be periodic in z or φ, since we use a new method incorporating a fast segmented convolution (chapter 3) in order to solve (2.123), and since the method is equivalent to Rayleigh-Sommerfeld diffraction (see chapter 4), which is a more general approach than the simplifications made in [Kuz97] that lead to Fresnel diffraction with a Fresnel parameter $N_F > 1$ [Mic98, Wie95]. Therefore the perturbation and the field structure can be of arbitrary shape as long as the perturbation is smooth in order for the phase corrector model to be applicable.

One could also start from the decomposition of a monochromatic spherical wave into cylindrical waves from [GR94]:

$$\frac{e^{-j\alpha\sqrt{r^2+z^2}}}{\sqrt{r^2+z^2}} = \frac{1}{2j} \int\limits_{-\infty}^{\infty} e^{-jtz} H_0^{(2)}(r\sqrt{\alpha^2 - t^2})dt \tag{2.126}$$

with $\alpha \equiv k, t \equiv k_z, r = |\vec{\varrho} - \vec{\varrho}\,'|$ and the addition theorem of cylindrical functions [AS64]

$$H_0^{(2)}(k_\rho |\vec{\varrho} - \vec{\varrho}\,'|) = \sum_{m=-\infty}^{\infty} H_m^{(2)}(k_\rho\rho')J_m(k_\rho\rho)e^{jm(\phi-\phi')} \quad \rho < \rho' \tag{2.127}$$

$$= \sum_{m=-\infty}^{\infty} H_m^{(2)}(k_\rho\rho)J_m(k_\rho\rho')e^{jm(\phi-\phi')} \quad \rho' < \rho \tag{2.128}$$

and get for $\rho < \rho'$:

$$\frac{e^{-jk|\vec{r}-\vec{r}'|}}{4\pi\,|\vec{r}-\vec{r}'|} = \frac{1}{8\pi j} \int\limits_{-\infty}^{\infty} \sum_{m=-\infty}^{\infty} H_m^{(2)}(k_\rho\rho')J_m(k_\rho\rho)e^{jm(\phi-\phi')}e^{-jk_z(z-z')}dk_z$$

$$(2.129)$$

Substituting this into the integral equation formalism will yield a related formulation for the cylindrical wave decomposition as shown in [LV06]. [LV06] also discusses the limits of the scalar theory also known as the physical optics approximation. Equation (2.126) is the cylindrical wave equivalent to Weyl's well known formula [Wey19] of plane wave decomposition of a monochromatic spherical wave.

2.3.5 Born's approximation

In order to solve the integral equations (2.111), (2.112) and (2.123) iteratively, we use a Born approximation of higher order [BW09]. The higher order is necessary due to the multiple reflections inside the launcher corresponding to multiple diffraction and scattering[7]. This simplest form of a series solution in diffraction theory corresponds to a Liouville-Neumann series which is a successive approximation method. The procedure is as follows:

- Start from an initial guess for the potential u_0, usually the initial incident field.

- Apply the operator to u_0 resulting in u_1

- For $n \geq 1$:
  Apply the operator to u_n resulting in u_{n+1}
  Calculate the residual error (see below)
  Stop when the residual error is small enough

---

[7]Scattering results from the phase corrector

In equation form this gives:

$$u_n = \underbrace{\overbrace{u_0 + \mathcal{L}u_0}^{u_2} + \mathcal{L}\mathcal{L}u_0}_{u_1} + \mathcal{L}\mathcal{L}\mathcal{L}u_0 + \ldots + \underbrace{\mathcal{L}\ldots\mathcal{L}}_{\text{n times } \mathcal{L}} u_0 \qquad (2.130)$$

where $\mathcal{L}$ is either one of the before mentioned operators.

One form of a residual is the root-mean-square (R.M.S.) change in u from one iteration to the next [Cát95]:

$$e_n^{r.m.s.} = \frac{\langle r_n, r_n \rangle^{1/2}}{\langle u_n, u_n \rangle^{1/2}} \qquad (2.131)$$

where[8]

$$\begin{aligned} r_n &= u_n - u_{n-1} \\ \langle f, g \rangle &= \iint_S f \cdot g^* dS \end{aligned} \qquad (2.132)$$

The optimum residual is $e_n^{r.m.s.} = 0$, due to the usually slow convergence of this series solution, a value of $e_n^{r.m.s.}$ in the low percent region is acceptable, due to the negligible visual change in the field solution, as demonstrated in sections 4.1.1-4.1.3. The numerical aspects of this approximation and the calculation of the integral equations aforementioned are discussed in chapter 3.

2.4 Electric Field Integral Equation (EFIE)

The final method, discussed here, used to calculate the electromagnetic field inside an oversized waveguide, is the Electric Field Integral Equation (EFIE). This equation gives a full three dimensional solution for both field vectors $\vec{H}$ and $\vec{E}$. This solution will be taken as the reference solution in the comparison later done in chapter 4, using the code described in [Nei04]. The derivation is based

---

[8]Equation (2.132) can easily be identified as the inner product of two functions f and g.

on [PRM98]. We start by expressing the total fields $\vec{E}^T$, $\vec{H}^T$ as superposition of incident $\vec{E}^I$, $\vec{H}^I$ and scattered fields $\vec{E}^S$, $\vec{H}^S$, that is:

$$\vec{H}^T = \vec{H}^I + \vec{H}^S \tag{2.133}$$

$$\vec{E}^T = \vec{E}^I + \vec{E}^S \tag{2.134}$$

$\vec{E}^I$ and $\vec{H}^I$ are the fields without the presence of the scatterer. Equations (2.22) and (2.23), relate the radiated scattered fields to the magnetic vector potential $\vec{A}$:

$$\vec{H}^S = \frac{1}{\mu} \nabla \times \vec{A} \tag{2.135}$$

$$\vec{E}^S = -j\omega\vec{A} - j\frac{\nabla(\nabla \cdot \vec{A})}{\omega\mu\varepsilon} \tag{2.136}$$

For the known incident fields $\vec{E}^I$ and $\vec{H}^I$, the equation for $\vec{A}$ has the homogeneous form as in (2.19):

$$\nabla^2 \vec{A} + k^2 \vec{A} = 0$$

In order to find the magnetic vector potential $\vec{A}$, related to the scattered part of the field $\vec{E}^S$ and $\vec{H}^S$, we have to solve the inhomogeneous form of equation (2.19):

$$\nabla^2 \vec{A} + k^2 \vec{A} = -\mu\vec{J} \tag{2.137}$$

The same formalism, but different integration contour as used for the scalar Helmholtz equation in section 2.3, leads to the solution [vB07]:

$$\vec{A}(\vec{r}) = \frac{\mu}{4\pi} \iiint\limits_V \vec{J}(\vec{r}')\frac{e^{-jk|\vec{r}-\vec{r}'|}}{|\vec{r}-\vec{r}'|}dV' \tag{2.138}$$

In this equation the unknown equivalent current $\vec{J}(\vec{r}')$ has to be found. To find the current we substitute (2.138) into (2.136) and get:

$$\vec{E}^S(\vec{r}) = \frac{-j\omega\mu}{4\pi} \iiint\limits_V \vec{J}(\vec{r}')\frac{e^{-jk|\vec{r}-\vec{r}'|}}{|\vec{r}-\vec{r}'|}dV'$$

$$- \frac{j\mu}{4\pi\omega\mu\varepsilon}\nabla\left(\nabla \cdot \iiint\limits_V \vec{J}(\vec{r}')\frac{e^{-jk|\vec{r}-\vec{r}'|}}{|\vec{r}-\vec{r}'|}dV'\right) \tag{2.139}$$

Since the problems only involve perfect electric conductor (P.E.C), the involved equivalent current density $\vec{J}(\vec{r}')$ is a surface current density $\vec{J}_S(\vec{r}')$ and the integration is done over a surface not over a volume. So the integrals simplify to surface integrals instead of volume integrals and we get, with further simplifications and as before $d = |\vec{r} - \vec{r}'|$:

$$\vec{E}^S(\vec{r}) = \frac{-j\omega\mu}{4\pi} \left(\iint_S \vec{J}_S(\vec{r}') \frac{e^{-jkd}}{d} dS' \right.$$
$$\left. + \frac{1}{k^2} \nabla \left(\nabla \cdot \iint_S \vec{J}_S(\vec{r}') \frac{e^{-jkd}}{d} dS' \right) \right) \tag{2.140}$$

Inserting this into (2.134) gives:

$$\vec{E}^T(\vec{r}) = \vec{E}^I(\vec{r}) + \frac{-j\omega\mu}{4\pi} \left(\iint_S \vec{J}_S(\vec{r}') \frac{e^{-jkd}}{d} dS' \right.$$
$$\left. + \frac{1}{k^2} \nabla \left(\nabla \cdot \iint_S \vec{J}_S(\vec{r}') \frac{e^{-jkd}}{d} dS' \right) \right) \tag{2.141}$$

Restricting the observation point to the scatterer ($\vec{r} = \vec{r}_S$) and applying the boundary condition

$$\vec{n} \times \vec{E}^T = \vec{0} \quad \text{at} \quad \vec{r} = \vec{r}_S \tag{2.142}$$

on the scatterer, gives the Electric Field Integral Equation

$$\vec{n} \times \vec{E}^I = -\vec{n} \times \vec{E}^S \tag{2.143}$$

$$= \vec{n} \times \frac{j\omega\mu}{4\pi} \left(\iint_S \vec{J}_S(\vec{r}') \frac{e^{-jkd}}{d} dS' \right.$$
$$\left. + \frac{1}{k^2} \nabla \left(\nabla \cdot \iint_S \vec{J}_S(\vec{r}') \frac{e^{-jkd}}{d} dS' \right) \right) \tag{2.144}$$

This equation relates the unknown surface current density $\vec{J}_s$ induced on the scatterer ($\vec{r} = \vec{r}_S$) to the tangent known incident electric field $\vec{n} \times \vec{E}^I$. It can be solved iteratively for $\vec{J}_S$ using the Method of Moments (MoM), once the current density is found the resulting scattered fields can be calculated using (2.140) directly or (2.135) after substituting $\vec{A}$. The Method of Moments and the numerical aspects of the solution of the EFIE will be discussed in the subsequent chapter.

3 Algorithms and numerical aspects

After deriving the integral equations for the four methods used, we now discuss the numerical aspects of solving these equations. The three scalar equations can be identified as convolution integrals. We will therefore use the convolution theorem of the Fourier Transform to obtain a fast algorithm, employing the very well known Fast Fourier Transform (FFT) [Bri88], for solving the scalar integral equations. This necessitates the restriction to a constant average radius, in order to perform the Fourier Transform in two dimensions. One possible solution for launchers with tapered average radius is introduced in section 3.5. The Electric Field Integral Equation is solved using the Method of Moments incorporating a fast algorithm for matrix-vector multiplication implemented in a commercial code [Nei04]. This last approach is the most time consuming, due to solving the problem in all three dimensions, as opposed to the two dimensional solution for the scalar integral equations.

The chapter starts with the convolution integrals and their solution using the Fourier Transform and their discrete versions. Then, the chapter continues with a short explanation of the segmented convolution. Subsequently the numerical aspects of each integral equation and a segmented convolution algorithm for two of the scalar integral equations are discussed, followed by an expansion of the segmented convolution algorithm in order to calculate tapered average radius launchers. The chapter concludes with a summary of the algorithms used for solving the Electric Field Integral Equation.

3.1 Convolution integrals and the Fourier Transform

The multidimensional convolution integral of two functions, s_1 and g, is defined as [Bam89]:

$$s_2(\vec{x}) = \int\limits_{-\infty}^{\infty} \cdots \int\limits_{-\infty}^{\infty} s_1(\vec{x}') \cdot g(\vec{x} - \vec{x}') d^n\vec{x}' = s_1(\vec{x}) \circledast g(\vec{x}) \qquad (3.1)$$

Additionally, we define[1] the multidimensional Fourier Transform as [Bam89]:

$$S_2(\vec{f}) = \int\limits_{-\infty}^{\infty} \cdots \int\limits_{-\infty}^{\infty} s_2(\vec{x}) \cdot e^{-j2\pi\vec{f}\cdot\vec{x}} d^n\vec{x} = \mathcal{F}\{s_2(\vec{x})\} \qquad (3.2)$$

and the inverse Fourier Transform as:

$$s_2(\vec{x}) = \int\limits_{-\infty}^{\infty} \cdots \int\limits_{-\infty}^{\infty} S_2(\vec{f}) \cdot e^{j2\pi\vec{x}\cdot\vec{f}} d^n\vec{f} = \mathcal{F}^{-1}\{S_2(\vec{f})\} \qquad (3.3)$$

Substituting these two equations for each function s_2, s_1 and g into equation (3.1) gives after some calculation the well known convolution theorem of the Fourier Transform given as:

$$S_2(\vec{f}) = S_1(\vec{f}) \cdot G(\vec{f}) \qquad (3.4)$$

Performing the inverse Fourier Transform on $S_2(\vec{f})$ gives the convolution result $s_2(\vec{x})$.

$$s_2(\vec{x}) = \mathcal{F}^{-1}\{S_1(\vec{f}) \cdot G(\vec{f})\} = \mathcal{F}^{-1}\{\mathcal{F}\{s_1(\vec{x})\} \cdot \mathcal{F}\{g(\vec{x})\}\} = s_1(\vec{x}) \circledast g(\vec{x})$$
$$(3.5)$$

That is the convolution s_2 of the two functions s_1 and g, can also be calculated by multiplying the spectra of the two functions and then taking the inverse Fourier

---

[1] The factor of $\frac{1}{2\pi}$ in front of the inverse transform, found in some textbooks, can be suppressed by adding a factor of 2π in the exponent of the transform.

Transform of this product[2]. Now considering the given integral equations, we can identify for (2.111):

$$s_1 = u^{(P)}(R_0, \varphi', z') \tag{3.6}$$

$$g = g^{(P)} = \left(\frac{e^{-jkr_0}}{r_0^2}\right)\left(jk + \frac{1}{r_0}\right)\cdot 2\cdot R_0^2\cdot \sin^2\left(\frac{\varphi - \varphi'}{2}\right) \tag{3.7}$$

and for (2.112):

$$s_1 = u^{(D)}(R_0, \varphi', z') \tag{3.8}$$

$$g = g^{(D)} = \left(\frac{e^{-jkr_0}}{r_0^2}\right)\left(jk + \frac{1}{r_0}\right)\cdot 4\cdot R_0^2\cdot \sin^2\left(\frac{\varphi - \varphi'}{2}\right) \tag{3.9}$$

Now we can calculate the desired convolutions by means of the Fourier Transform. For the CWD (2.120), the integral is already in the form for the convolution theorem and we get:

$$\psi^{(2)}(\rho, \varphi, z) = \frac{1}{4\pi^2} \overbrace{\sum_{-\infty}^{\infty} \int_{-\infty}^{\infty} \underbrace{\frac{H_m^{(2)}\left(\rho\sqrt{k^2 - k_z^2}\right)}{H_m^{(1)}\left(R_0\sqrt{k^2 - k_z^2}\right)}}^{\mathcal{F}\{g\}}}^{\mathcal{F}^{-1}\{S_1\cdot G\}}$$

$$\underbrace{\int_0^{2\pi}\int_{-\infty}^{\infty} \psi^{(1)}(R_0, \varphi', z')\cdot e^{-jk_z z'}e^{-jm\varphi'}\,dz'\,d\varphi'}_{\mathcal{F}\{s_1\}}\underbrace{e^{jm\varphi}e^{jk_z z}dk_z}_{\mathcal{F}^{-1}\{S_1\cdot G\}} \tag{3.10}$$

In order to numerically calculate the convolutions with a computer program, the functions have to be sampled. A sampled form of the Fourier Transform is the Discrete Fourier Transform. Under certain conditions (section 3.1.1), the Fourier Transform of a function can be approximated by means of the Discrete Fourier Transform (DFT). The DFT has a very fast, well known implementation, called the Fast Fourier Transform (FFT). This makes it possible to calculate the three scalar convolution integrals in a very short time. The numerical cost is usually

---

[2]A summary on the Fourier Transform in English in the context of diffraction can be found in [Goo05]

determined by the amount of necessary operations. For the regular algorithm of the transform, the number of operations is about $\mathcal{O}(N^2)$, for N elements to be transformed. The FFT implementation needs for the same amount of elements $\mathcal{O}(N\ log_2 N)$ operations. This example shows the possible speed up we achieve using the FFT.

3.1.1 Discrete Fourier Transform and cyclic convolution

We will now discuss the conditions necessary to calculate the Fourier Transform by means of the DFT. We will discuss this in one dimension only, since the expansion to two or more dimensions is straight forward. See [Bri88] for an in depth treatment of DFT, FFT and its applications. The discussion is illustrated using figure 3.1. The left side of the figure shows the space domain and the right side shows the spatial frequency domain.

In one dimension, the sampling of a continuous function $h(x)$ (Fig. 3.1(a)) is done by multiplying it with a train of delta functions (Fig. 3.1(b)).

$$
\begin{aligned}
h(x) \cdot \mathrm{train}_{\delta_x} &= h(x) \sum_{q=-\infty}^{\infty} \delta(x - q\Delta x) \\
&= \sum_{q=-\infty}^{\infty} h(q\Delta x)\delta(x - q\Delta x) \qquad (3.11)
\end{aligned}
$$

Δx is the sampling interval. For the transform this results in a periodic repetition of the spectrum as shown in Figure 3.1(c). The occurring overlap in the frequency domain is known as aliasing. In order to avoid aliasing, we get the well known sampling condition

$$
\Delta x \leq \frac{1}{2 f_{max}}, \qquad (3.12)
$$

where f_{max} denotes the maximum spatial frequency of the band-limited function $h(x)$.

Due to the limited memory in a computer, the function has to be truncated after N samples. This corresponds to the multiplication with a rectangle window

(Fig.3.1(d)) and (3.11) yields

$$h(x) \cdot \text{train}_{\delta_x} \cdot \text{rect}(x) = \left[\sum_{q=-\infty}^{\infty} h(q\Delta x)\delta(x - q\Delta x) \right] \cdot \text{rect(x)}$$

$$= \sum_{q=0}^{N-1} h(q\Delta x)\delta(x - q\Delta x). \qquad (3.13)$$

For the Fourier Transform this results in an added ripple to the periodic spectrum, shown in figure 3.1(e). This ripple is due to the convolution of the periodic spectrum with the spectrum of the rectangle. If possible, the truncation Period T should be a multiple integer of Δx, in order to avoid the effect known as leakage [Bri88]. Additionally, analog to the space domain, the spectrum is sampled and is therefore multiplied by a train of delta functions (Fig.3.1(f)) in the frequency domain.

$$\tilde{h}(x) = h(x) \cdot \text{train}_{\delta_x} \cdot \text{rect}(x) \circledast \text{train}_{\delta_f}$$

$$= \left[\sum_{q=0}^{N-1} h(q\Delta x)\delta(x - q\Delta x) \right] \circledast \text{train}_{\delta_f}$$

$$= T \sum_{p=-\infty}^{\infty} \left[\sum_{q=0}^{N-1} h(q\Delta x)\delta(x - q\Delta x - pT) \right] \qquad (3.14)$$

The resulting function $\tilde{h}(x)$ is for N samples an approximation to the function $h(x)$. Sampling a continuous function results in periodicity in space and frequency domain as illustrated in figure 3.1(g). So the discrete Fourier Transform and its inverse can be applied when approximating the functions in periodic form. The discrete Fourier Transform over N samples is defined as:

$$\tilde{H}\left(\frac{n}{N\Delta x}\right) = \sum_{q=0}^{N-1} h(q\Delta x)e^{-j2\pi nq/N} \quad n = 0, 1, \cdots, N-1 \qquad (3.15)$$

and

$$\Delta x \cdot \tilde{H}\left(\frac{n}{N\Delta x}\right) = \Delta x \cdot \sum_{q=0}^{N-1} h(q\Delta x)e^{-j2\pi nq/N}$$

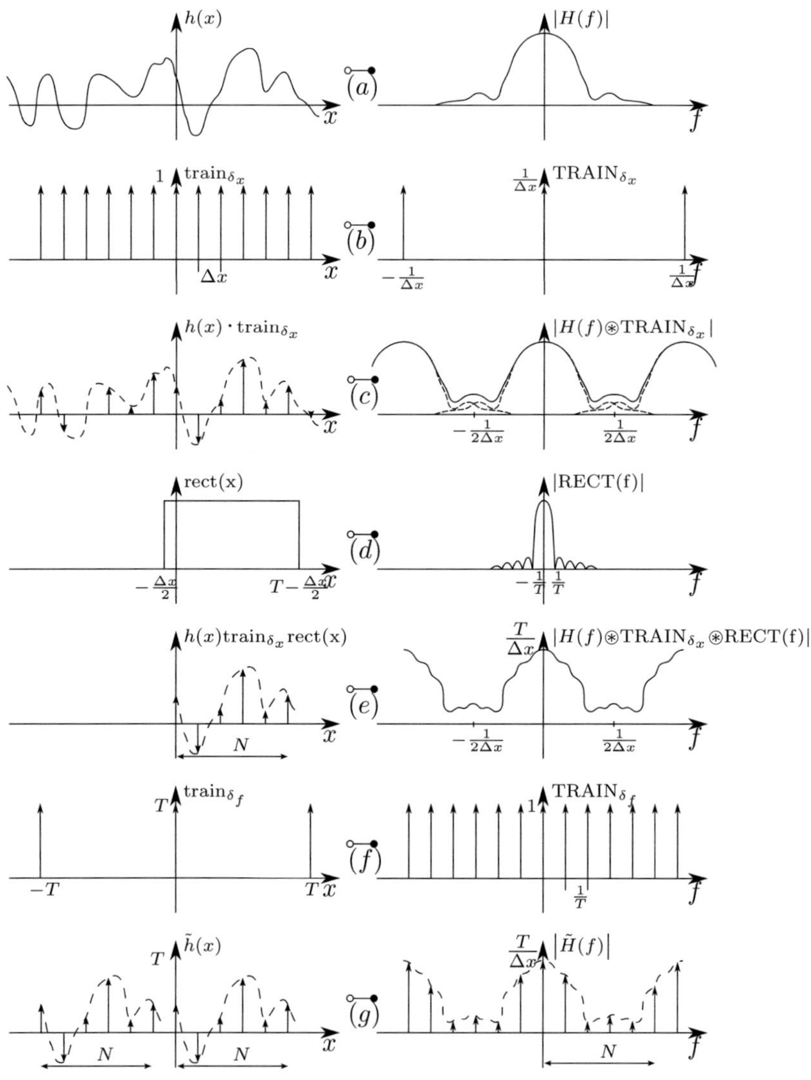

Figure 3.1: Effects on the continuous Fourier Transform due to sampling

is an approximation to the continuous Fourier Transform $H(f)$.

The inverse discrete Fourier Transform is defined as:

$$h(q\Delta x) = \frac{1}{N} \sum_{n=0}^{N-1} \tilde{H}\left(\frac{n}{N\Delta x}\right) e^{j2\pi nq/N} \quad q = 0, 1, \cdots, N-1 \qquad (3.16)$$

and

$$\frac{1}{\Delta x} h(q\Delta x) = \frac{1}{\Delta x N} \sum_{n=0}^{N-1} \tilde{H}\left(\frac{n}{N\Delta x}\right) e^{j2\pi nq/N}$$

is an approximation to the inverse continuous Fourier Transform.

Due to the periodicity of the functions in space and frequency domain due to sampling, the resulting discrete convolution

$$\begin{aligned}
y(i\Delta x) &= h(i\Delta x) \circledast f(i\Delta x) \\
&= \sum_{q=0}^{N-1} h(q\Delta x) f((i-q)\Delta x) \\
&= \mathcal{F}_d^{-1}\left\{ \tilde{H}\left(\frac{n}{N\Delta x}\right) \tilde{F}\left(\frac{n}{N\Delta x}\right) \right\} \qquad (3.17)
\end{aligned}$$

is a cyclic convolution[3]. In other words the convolution wraps around at the ends. In order to perform the aperiodic convolution necessary to calculate the diffraction inside an oversized waveguide, one has to find a solution in which the wrap-around effect is negligible. This is treated in the following section.

3.1.2 Aperiodic and segmented convolution

In order for the convolution (3.17) of two sampled functions to be aperiodic, one has to choose the resulting sample number N large enough and zero pad the initial sampled functions up to the sample number N. We will illustrate the necessary sample number N using an example. Figure 3.2 illustrates two sampled and zero padded example functions $h(i\Delta x)$, $f(i\Delta x)$ and the resulting convolution $y(i\Delta x)$ $(\Delta x = 1)$. Considering the figure, the necessary sample number N can be determined from the initial samples of $h(i\Delta x)$ and $f(i\Delta x)$. Let N_h be the

---

[3] $\mathcal{F}_d$ denotes the discrete version of the transform

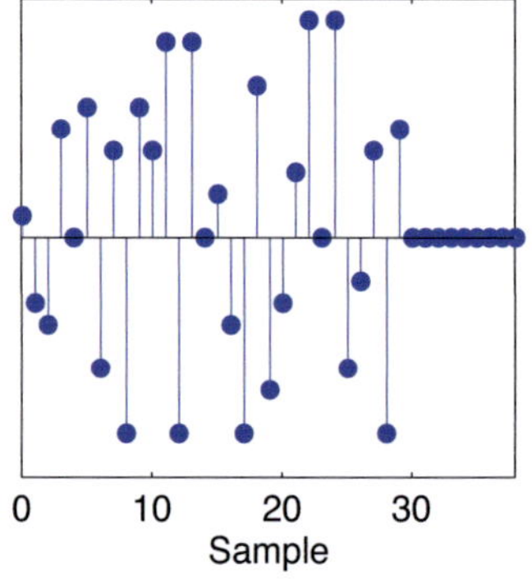

(a) Sampled and zero padded example function $h(i)$, $\Delta x = 1$

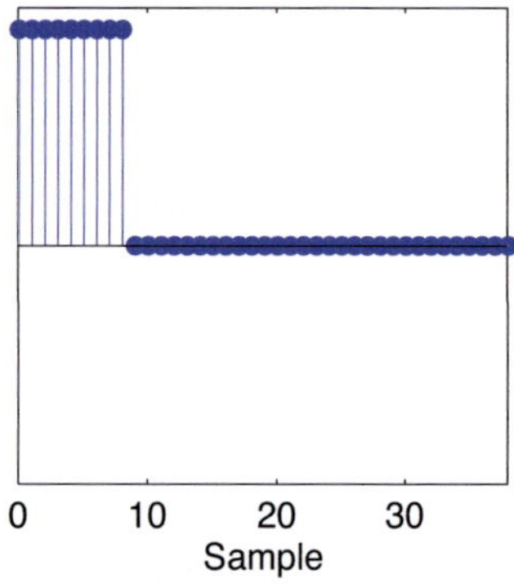

(b) Sampled and zero padded example function $f(i)$, $\Delta x = 1$

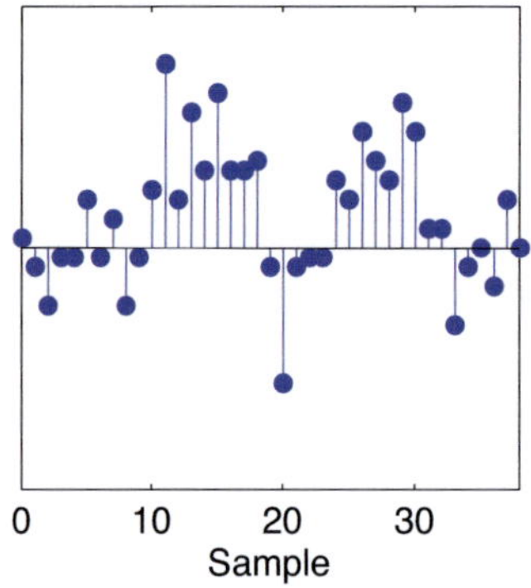

(c) Resulting convolution $y(i)$, $\Delta x = 1$

Figure 3.2: Example of a one dimensional aperiodic convolution

number of samples of $h(i\Delta x)$ and N_f the number of samples of $f(i\Delta x)$, then the resulting convolution has $N = N_h + N_f - 1$ samples. So in order to perform the aperiodic convolution using the DFT each of the functions has to be zero padded to contain N samples.

If one of the two functions to be convolved has substantially more samples than the other[4], it is possible, due to the linear property of the convolution operation, to split the larger function into smaller segments. Then convolve each segment with the second function and sum the results of each smaller convolution to give the result of the complete convolution. This procedure is called segmented or sectioned convolution [Bri88]. There are two common types of the segmented convolution, the overlap-add and the overlap-save type. Only the overlap-add segmented convolution will be considered here. An adapted form called the weighted overlap-add segmentation [Rab78] will be used for the calculation of diffraction in tapered launchers (section 3.5).

Segmentation example

The aforementioned overlap-add segmentation is illustrated using the example convolution from figure 3.2 (again $\Delta x = 1$). The first step is the segmentation of the function $h(i)$. In this example 3 segments are chosen. For a good choice of sample amount N for the segmentation, see [Bri88]. The first segment $h_1(i)$ corresponds to samples 0-9 (Fig. 3.3(a)), the second segment $h_2(i)$ corresponds to samples $10 - 19$ (Fig. 3.3(b)) and the last segment $h_3(i)$ corresponds to samples $20 - 29$ (Fig. 3.3(c)). The segments are all shifted to the origin before being zero padded. In this example, the segments and the function $f(i)$ (Fig. 3.3(d)) are zero padded to a sample length $N = N_{h_{1,2,3}} + N_f - 1 = 19$. The results $y_1(i)$, $y_2(i)$ and $y_3(i)$ of the convolution of each segment with $f(i)$ are shown in figure 3.4. The results are added with a given overlap of 9 samples (Fig. 3.5). Considering figure 3.5(a) the overlap for the first two segments starts at sample 10. The blue colored points are samples from the first segment, whereas the green color depicts samples from the second segment. The dashed magenta color corresponds to the sum of the segments. The parts without overlap are in magenta color, since the sum corresponds to the samples of the corresponding segment. For the second

---

[4]In digital filtering, the impulse response of the filter is usually much shorter than the input signal being filtered

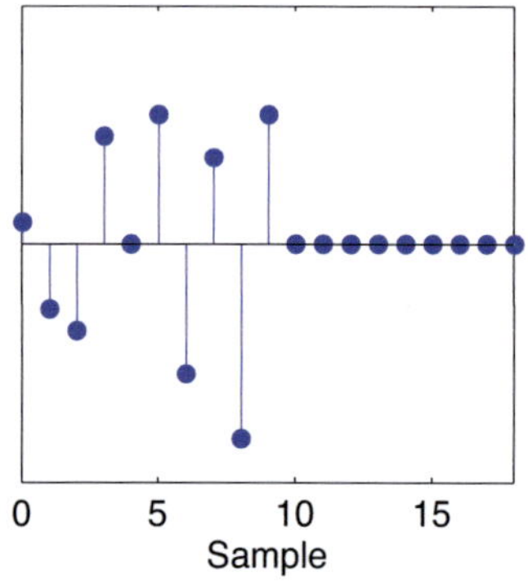

(a) Sampled and zero padded segment $h_1(i)$

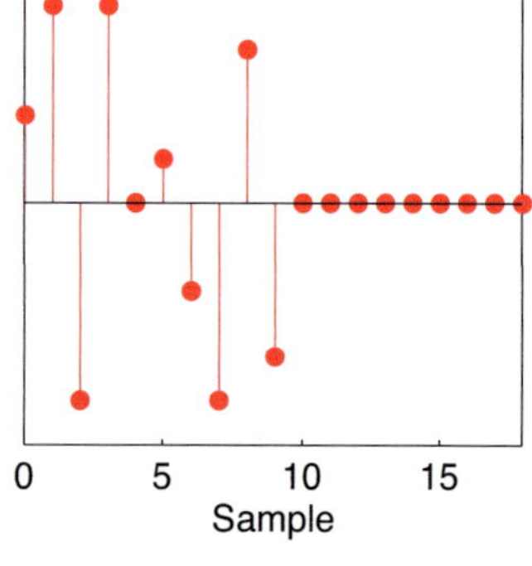

(b) Sampled and zero padded segment $h_2(i)$

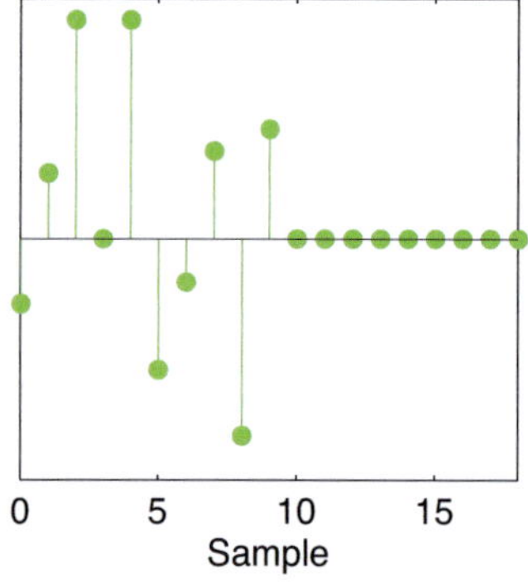

(c) Sampled and zero padded segment $h_3(i)$

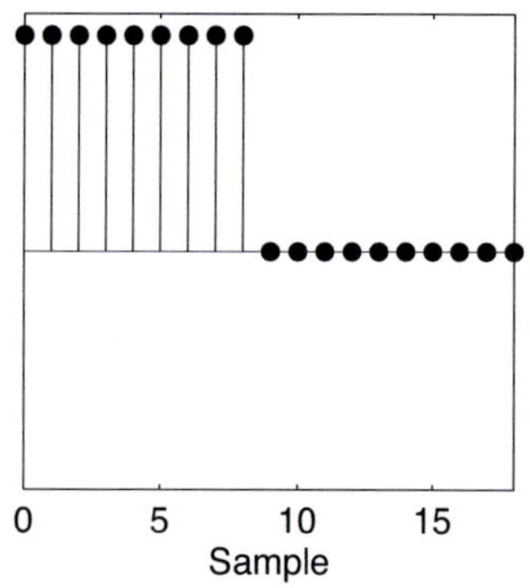

(d) Sampled and zero padded function $f(i)$

Figure 3.3: Zero padded segments $h_1(i)$, $h_2(i)$, $h_3(i)$ and $f(i)$ for the overlap-add segmented convolution

overlap, the result from the first overlap-add part is used and overlapped and added at the correct position (starting at sample 20) with the third segment (Fig. 3.5(b)). The magenta colored samples correspond to the result of the first overlap-add part, the red colored samples are the result of the third segment. The dashed black colored samples are the resulting sum of the segments and correspond to the final segmentation result shown in figure 3.5(c). Now comparing both results of the segmented convolution (Fig. 3.5(c)) and the non-segmented convolution (Fig. 3.5(d)) shows no difference. So both algorithms give identical results as expected.

The two dimensional form of this segmented convolution algorithm will be applied to the DSD and the CWD for fast computation of the integral equations. The segmentation algorithm is not applied to PSD, due to the existing implementation of this type of diffraction in the computer code [JTP+09] named TWLDO.

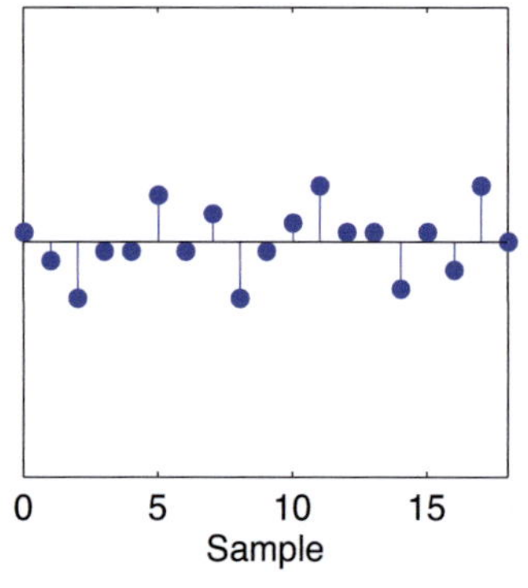

(a) Convolution result of first segment $y_1(i)$

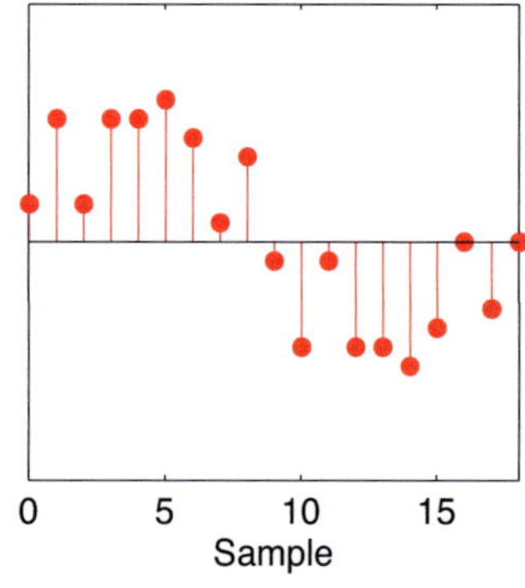

(b) Convolution result of second segment $y_2(i)$

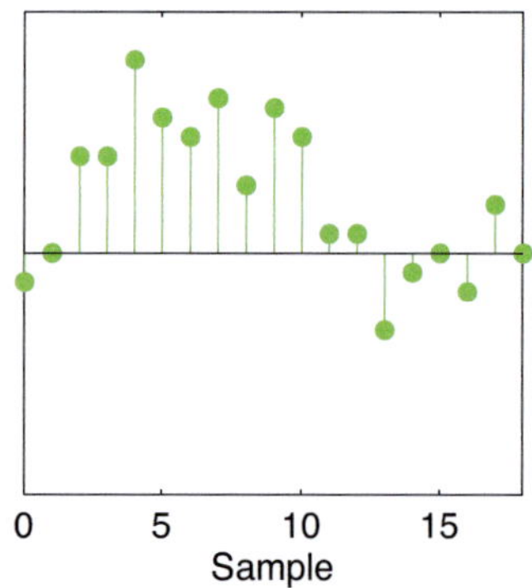

(c) Convolution result of third segment $y_3(i)$

Figure 3.4: Convolved segments $y_1(i)$, $y_2(i)$, $y_3(i)$ for the overlap-add segmented convolution

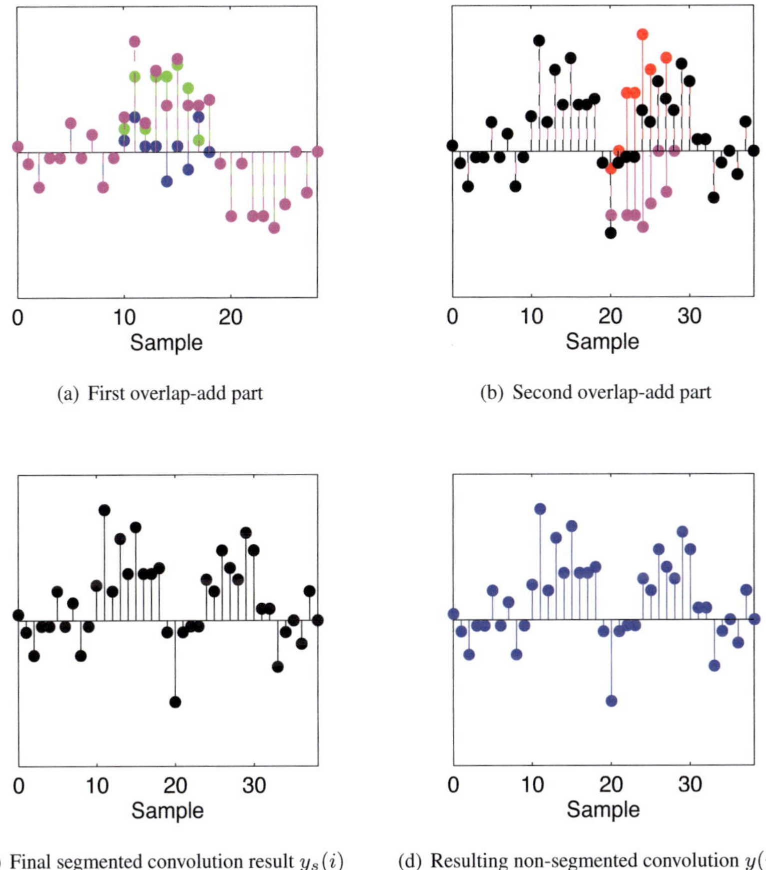

(a) First overlap-add part

(b) Second overlap-add part

(c) Final segmented convolution result $y_s(i)$

(d) Resulting non-segmented convolution $y(i)$ (from Fig. 3.2(c))

Figure 3.5: Overlap-add procedure of the individual segments and final convolution result

3.2 Numerical aspects for the PSD

Since the PSD has an existing implementation and the method has been described in detail in [JTP$^+$09], the numerical aspects of this method will only be discussed briefly. The focus of the discussion lies on the FFT of the sampled integral equation kernel $g^{(P)}$ and the length of the initial field distribution u_0. We repeat the integral equation (2.111), for better reference:

$$u^{(P)}(R_0, \varphi, z) = e^{-j2k_\rho \Delta R(\varphi,z)\sqrt{1-\left(\frac{m}{x'_{mn}}\right)^2}} \cdot \frac{1}{4\pi} \int\limits_{0}^{2\pi} \int\limits_{-\infty}^{L(\varphi)} u^{(P)}(R_0, \varphi', z,')$$

$$\cdot \left(\frac{e^{-jkr_0}}{r_0^2}\right) \cdot \left(jk + \frac{1}{r_0}\right) \cdot 2 \cdot R_0 \cdot \sin^2\left(\frac{\varphi - \varphi'}{2}\right) R_0 dz' d\varphi'$$

The sampled form of the kernel $g^{(P)}$ from equation 3.7 is given as:

$$g^{(P)}(p\Delta\varphi, q\Delta z) \;=\; \frac{1}{4\pi}\left(\frac{e^{-jk\Delta r_0}}{\Delta r_0^2}\right)\left(jk + \frac{1}{\Delta r_0}\right) \cdot$$

$$2 \cdot R_0^2 \cdot \sin^2\left(\frac{p\Delta\varphi}{2}\right) \tag{3.18}$$

with

$$\Delta r_0 = \sqrt{(q\Delta z)^2 + 4R_0^2 sin^2\left(\frac{p\Delta\varphi}{2}\right)}, \tag{3.19}$$

where $\Delta\varphi$ and Δz correspond to the discretization in φ and z. p and q are the index variables for the two dimensional array of size $P \times Q$. The singularity of $g^{(P)}$ at $\Delta r_0 = 0$ is implemented by sampling very close to the singularity. For visibility reasons, we disclaim the figures of magnitude and phase for a sampled example kernel $g^{(P)}$. Figure 3.6 shows the magnitude and Figure 3.8 the phase of the FFT of an example PSD kernel, with

$$m_{min} = -\frac{P}{2\Delta\varphi} \quad \text{and} \quad m_{max} = \frac{P}{2\Delta\varphi} - 1\,,$$

$$k_{z_{min}} = -\frac{Q}{2\Delta z} \quad \text{and} \quad k_{z_{max}} = \frac{Q}{2\Delta z} - 1.$$

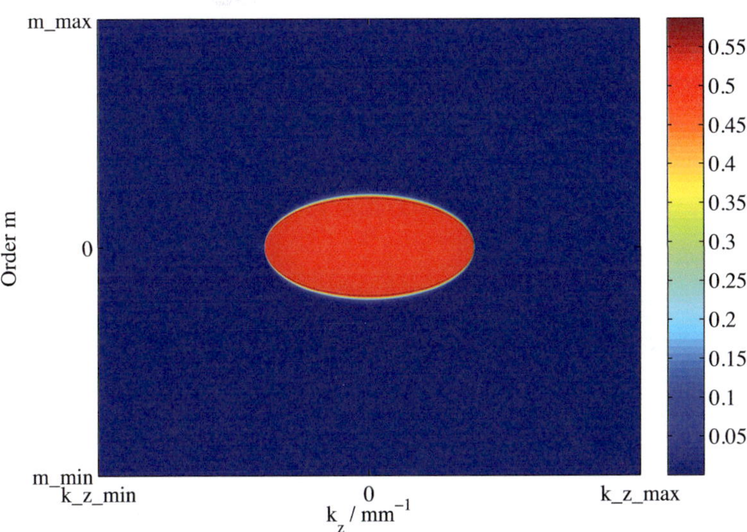

Figure 3.6: Magnitude of the FFT of an example PSD kernel

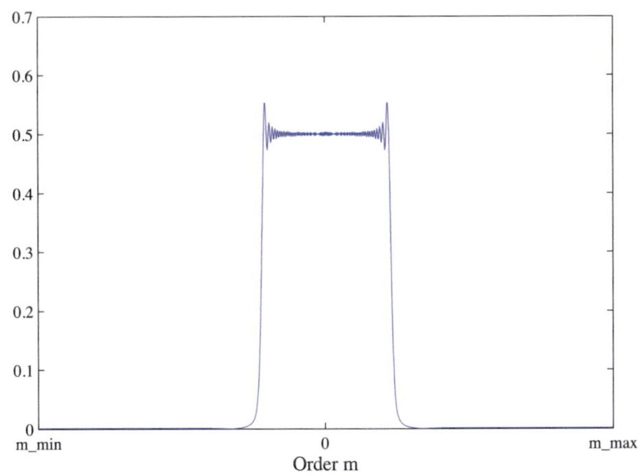

Figure 3.7: Magnitude of FFT over orders m at $k_z = 0$ showing the overshoot near the edge

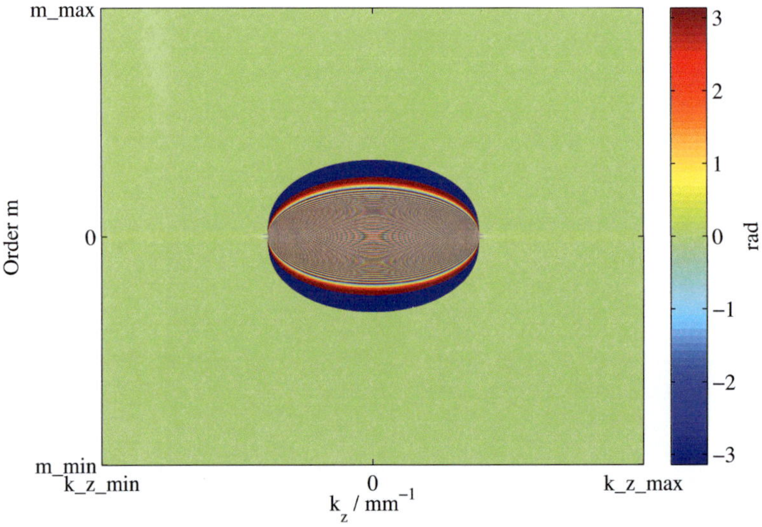

Figure 3.8: Phase of the FFT of an example PSD kernel

Using figures 3.6 and 3.8, the properties of the propagation can be illustrated very well: The FFT of the PSD kernel corresponds to a transfer function, known from linear system theory. This transfer function describes the propagation by one Brillouin length L_B in z-direction and has similarities with a two dimensional low pass filter[5]. That is, up to a certain order m and certain wave number k_z, we have a constant magnitude of $\left|\mathcal{F}_d\{g^{(P)}\}\right| \approx 0.5$, which after an overshoot[6] (Fig. 3.7) near the edge reduces to zero (Fig. 3.6). This constant magnitude exhibits a reduction of the spectral components of the field by a factor 2 in each iteration and does not correspond to the usually described loss-free transmission from one Brillouin zone to the next in an oversized waveguide [Kuz97]. Considering

---

[5]To avoid confusion, it should be noted that in terms of frequency f, a hollow waveguide corresponds to a high pass filter.

[6]resulting from Gibbs' phenomenon

figure 3.8, the phase of the transfer function shows a very rapid fluctuating pattern. This can be interpreted as follows: Each spectral component experiences an individual phase shift over the same geometrical distance.

When numerically evaluating (2.111) over a finite domain, the error compared to the evaluation to $-\infty$ has to be negligible. The error is negligible when limiting the domain to $-7L$ in negative z-direction, where L is the launcher length [JTP$^+$09]. By loosening the restriction on the error criteria, the domain can be reduced to $-3L$ [Jin11]. This results in a total length for the calculation domain of $4L$, when adding the launcher length to the domain.

The initial field u_0 for $-3L < z < L$ is given by

$$u_0^{(P)} = u_0(\vec{r}') = A_{mn} J_m(k_\rho R_0) e^{-jm\varphi'} e^{-jk_z z'} \tag{3.20}$$

and corresponds to the pure mode generated in the resonator of the gyrotron.

The resulting calculation domain is chosen in the order of $P \times Q = 1024 \times 4096$. This quite large calculation domain is necessary due to the transfer function used: The transfer function reduces the fields amplitude by half in each iteration (4.1.1). Since each iteration describes the propagation by one Brillouin length L_B, the "source field" u_0 has to provide "enough" field for the wave to reach the end of the calculation domain after typically $10 \sim 15$ iterations. The simulation results gained using this method are discussed and compared with the other presented methods in chapter 4.

3.3 Algorithm and numerical aspects of the DSD

Algorithm

The algorithm used to solve the integral equation (2.112) has been implemented in a C++ computer program named "Launcher". We illustrate the algorithm by applying it to a straight non-perturbed Vlasov type example launcher. Consequently $\Delta R(\varphi, z) = 0$ for all φ and z and equation (2.112) reduces to equation (2.105). The corresponding parameters of the example launcher are given in table 3.1 and coincide with the parameters from table 2.3 except for the launcher radius and the taper angle. Subsequently to the explanation of the algorithm and the

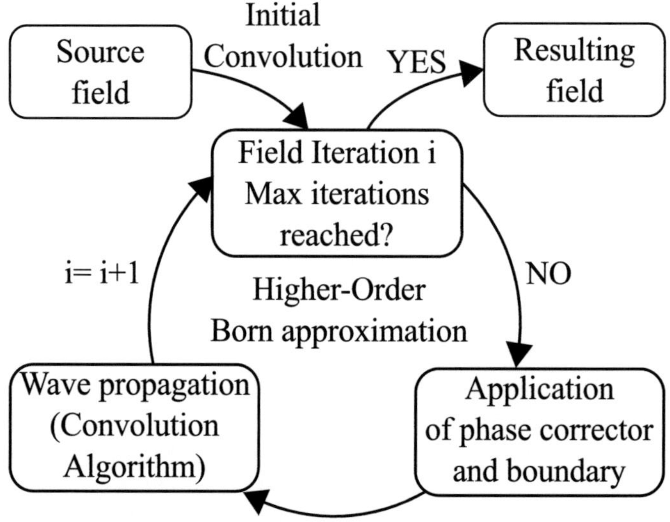

Figure 3.9: Iterative algorithm used to solve equation (2.112)

Parameter	Value
Operating mode	$TE_{34,19}$
Frequency f	170 GHz
Brillouin angle Ψ	65.29 °
Spread angle θ	71.14 °
Launcher radius R_0	32.5 mm
Taper angle $\tan\alpha$	0.000

Table 3.1: Parameters of the example launcher without perturbation

example calculation, we discuss the numerical aspects as well as the reasons for the chosen parameters of the example calculation. Figure 3.9 illustrates the used algorithm and can be explained as follows: Starting from a source field (Fig. 3.10) and performing an initial convolution, results in the $0th$-iteration of the field (Fig. 3.11). This result is the field propagating into the calculation domain and has to be added to each iteration as initial result. Therefore the initial source convolution, corresponding to the integration limit from $-\infty < z \leq 0$, has to be evaluated only once[7] and the calculation domain for the iterations can be restricted to the launcher length $(0 \leq z \leq L(\varphi))$ and additional domain for the radiated field. In the next step, we multiply the launcher boundary and the phase corrector, resulting from $\Delta R(\varphi, z)$, to the field. This reflects the outgoing wave from the waveguide wall and transforms it into the ingoing wave. The launcher boundary takes the φ dependence of the launcher into account and reflects only at the metal boundary. An example of the boundary is shown in figure 3.12. One can identify the cut of the launcher at $\varphi \approx 1.24\,\mathrm{rad}$, the overall length of $L \approx 291\,\mathrm{mm}$ and the beginning of the cut at $z = 220\,\mathrm{mm}$. Now the wave propagation is done by performing the convolution algorithm, shown in figure 3.16 and explained later on. This result gives the outgoing wave of the next iteration. Subsequently, we calculate and note the R.M.S. error from section 2.3.5 and start the next iteration. After reaching the necessary amount of iterations (~ 20), the algorithm stops. In order to save space, figure 3.13 shows only every second iteration of the example Vlasov launcher calculation. Figures 3.13(a) through 3.13(c) show the advancement of the field in the straight waveguide. Starting with figure 3.13(c), the influence (light blue colored form) of the cut of the launcher can be observed. The forming of the diffraction pattern can be observed starting in figure 3.13(e). The iterations 13 through 19 are disclaimed, since there is no noticeable visual change. The end result of the calculation is shown in figure 3.14(a). We can identify a two dimensional diffraction pattern at the end of the launcher in the zoomed part of the last iteration (Fig. 3.14(b)). This figure also shows reflection patterns forming in parallel to the helical part of the cut. The field that starts to radiate from the launcher is shown in figure 3.15. Now turning to figure 3.16, we explain the actual convolution algorithm used in each iteration i. We start be dividing the calculation domain into segments. Then for each segment in the domain, we zero

---

[7]The actual integration limits for the source convolution are discussed in the numerical aspects section

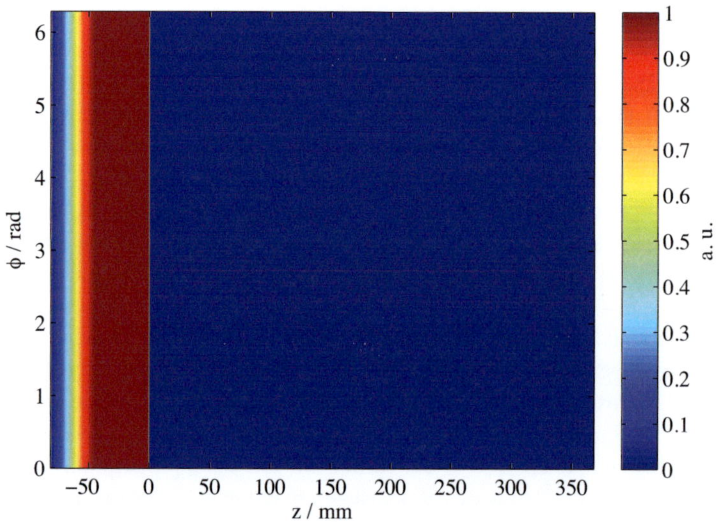

Figure 3.10: Magnitude of the source field used for initial convolution over φ- and z-direction @ $R_0 = 32.5\,\text{mm}$

pad the segment as illustrated in the segmented convolution example and calculate the FFT of the segment. The resulting spectrum is multiplied by the transfer function, resulting from the FFT of $g^{(D)}$. After the multiplication, we calculate the inverse FFT (IFFT) and add the result of this segment at the correct position to the existing results of already calculated segments. The convolution algorithm stops, when the last segment in the calculation domain has been evaluated. One of the segmented convolutions is shown exemplary in figure 3.17. Figure 3.17(a) shows the zero padded input segment, whereas figure 3.17(b) shows the resulting convolution. In this figure, it is shown how the field[8] propagates one Brillouin length $L_B \approx 28.3$ mm in one iteration. We now discuss the numerical aspects

---

[8]Of course, here we have only one segment of the field

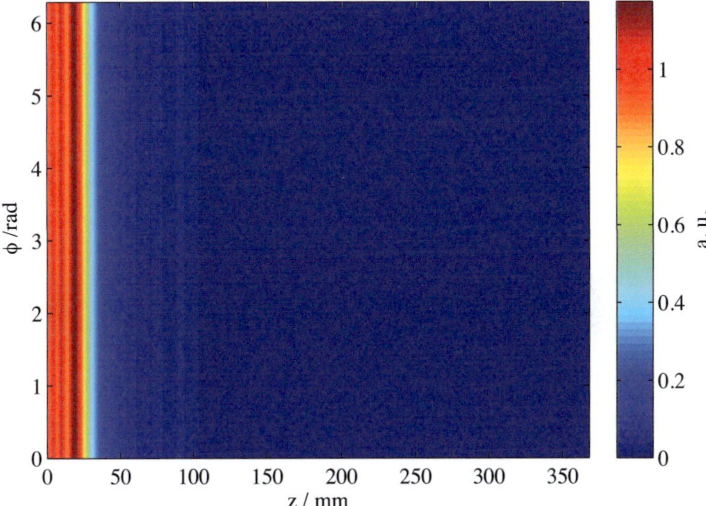

Figure 3.11: Magnitude of the outgoing field from the $0th$-iteration

and the chosen parameters of the integration kernel and the source field in the
following paragraph.

Numerical aspects

As before, we start by repeating the equation to be solved:

$$u^{(D)}(R_0, \varphi, z) = e^{-j2k_\rho \Delta R(\varphi,z)\sqrt{1-\left(\frac{m}{x'_{mn}}\right)^2}} \cdot \frac{1}{4\pi} \int\limits_{0}^{2\pi} \int\limits_{-\infty}^{L(\varphi)} u^{(D)}(R_0, \varphi', z,') \cdot$$

$$\left(\frac{e^{-jkr_0}}{r_0^2}\right)\left(jk + \frac{1}{r_0}\right) \cdot 4 \cdot R_0 \cdot \sin^2\left(\frac{\varphi - \varphi'}{2}\right) R_0 dz' d\varphi'$$

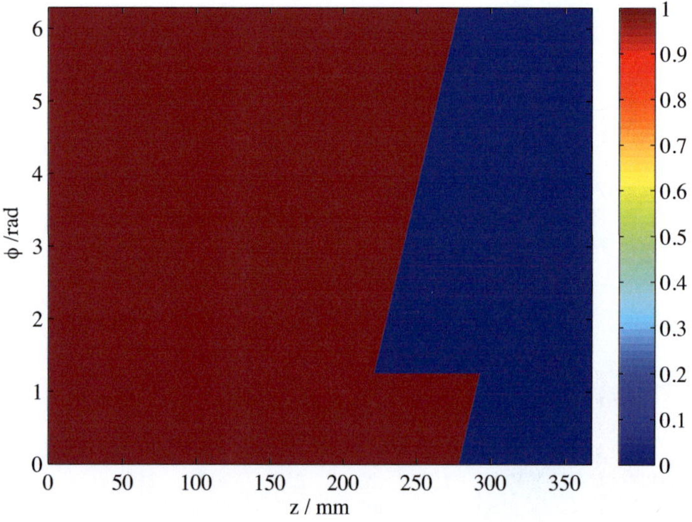

Figure 3.12: Launcher boundary (example)

The sampled form of the kernel $g^{(D)}$ from equation 3.9 is given as:

$$g^{(P)}(p\Delta\varphi, q\Delta z) = \frac{1}{4\pi}\left(\frac{e^{-jk\Delta r_0}}{\Delta r_0^2}\right)\left(jk + \frac{1}{\Delta r_0}\right) \cdot 4 \cdot R_0^2 \cdot \sin^2\left(\frac{p\Delta\varphi}{2}\right)$$

$$(3.21)$$

with the same notation as for the PSD kernel. The dipole source corresponds to a δ-point for $\Delta r_0 = 0$. We therefore use the integral value of a δ-function as the numerical implementation for the singularity [Bam89, vB07] :

$$\int_{S_\perp} \delta\left(r(\varphi, z)\right) dS_\perp = \frac{1}{|\nabla r(\varphi', z')|} \tag{3.22}$$

where $S_\perp$ is the path of integration intersecting perpendicular to the δ-function at $\varphi = \varphi'$ and $z = z'$.

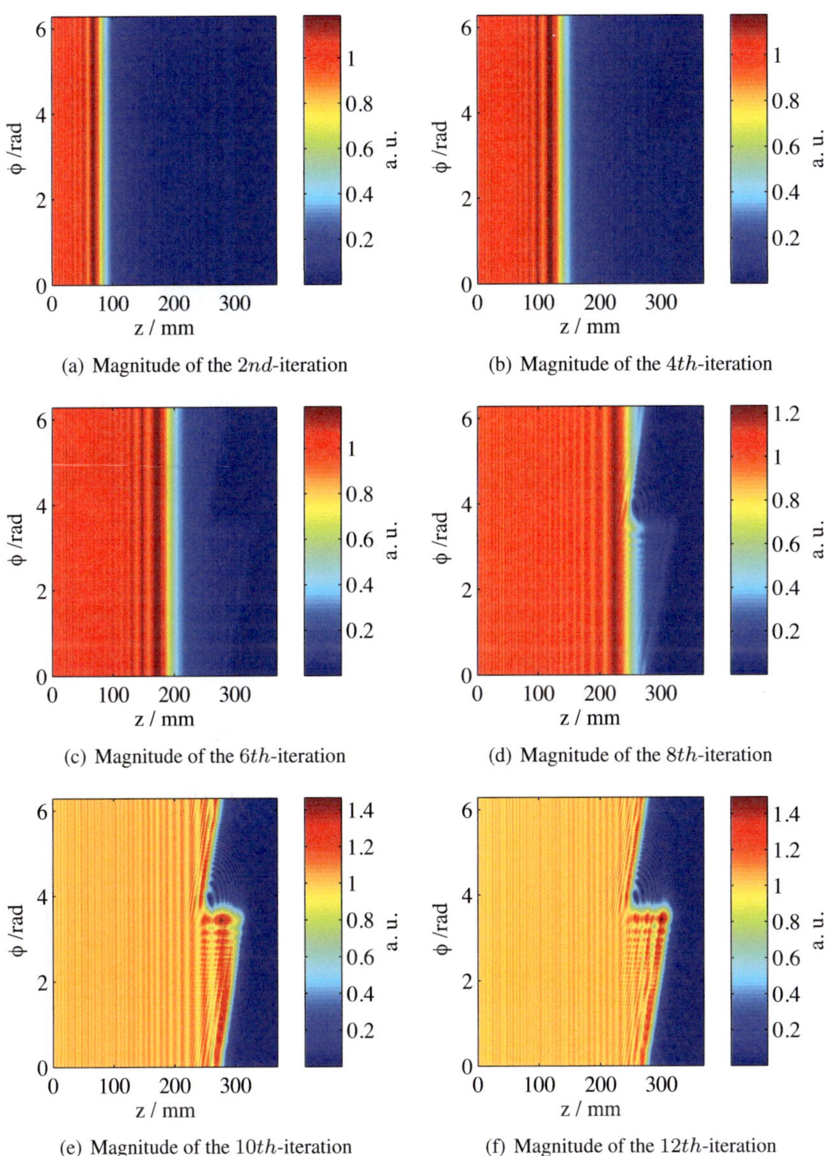

(a) Magnitude of the $2nd$-iteration

(b) Magnitude of the $4th$-iteration

(c) Magnitude of the $6th$-iteration

(d) Magnitude of the $8th$-iteration

(e) Magnitude of the $10th$-iteration

(f) Magnitude of the $12th$-iteration

Figure 3.13: Magnitudes of the outgoing field of iterations 2, 4, 6, 8, 10, 12

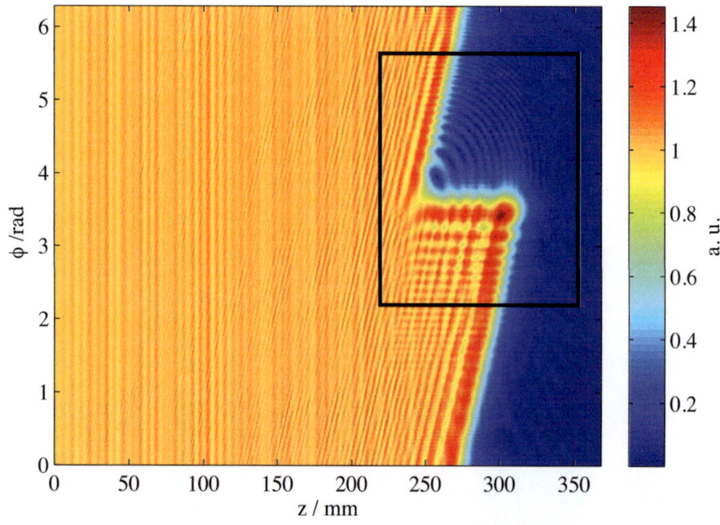

(a) Magnitude of the outgoing field the last iteration ($20th$)

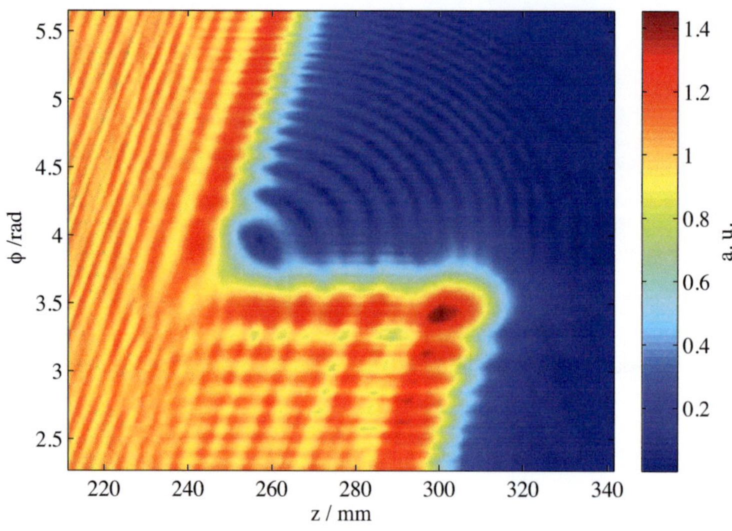

(b) Zoomed part of the last iteration ($20th$)

Figure 3.14: Result of the field of the last iteration ($20th$)

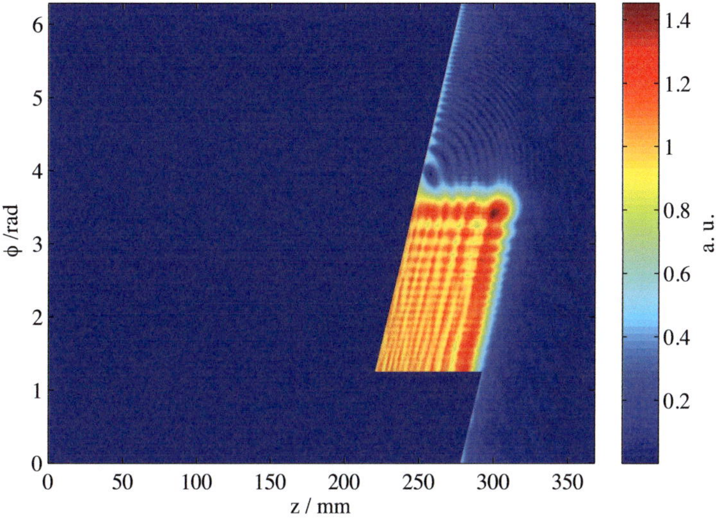

Figure 3.15: Magnitude of the field starting radiate from the launcher

Referring to geometrical optics in section 2.1.3, a ray travels the distance L_B in z-direction and the distance L_{trans} in φ-direction through the waveguide. The total distance L_{total} traveled can be determined, after substituting L_B and L_{trans} as follows:

$$L_{total} = \sqrt{L_B^2 + L_{trans}^2} = 2R_0 \frac{k}{k_\rho} \sqrt{1 - \left(\frac{m}{\chi_{mn}}\right)^2} \qquad (3.23)$$

If the ray is diffracted, the maximum of diffraction occurs when L_{total} is only diffracted in z-direction. Hence, we call L_{total} the diffraction length. Since diffraction can occur to both sides, that is in $\pm z$-direction, the extent of the kernel $g^{(D)}$ in z-direction is $2L_{total}$. Additionally, we have to zero pad the kernel in order to obtain an aperiodic convolution. In total, this results in the necessary

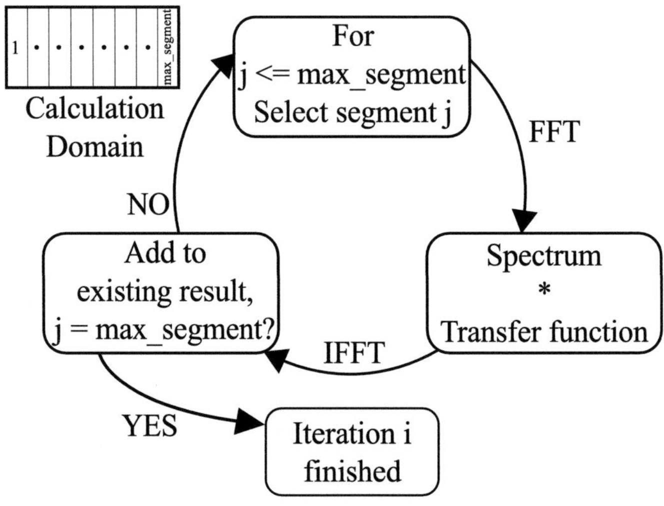

Figure 3.16: Convolution algorithm used to calculate each iteration i

number of samples

$$Q \geq \frac{4L_{total}}{\Delta z} \tag{3.24}$$

in z-direction. Since the convolution is circular in φ by nature of the problem, no zero padding or size adjustment is necessary in this direction. This reduces the calculation domain drastically in comparison with the implementation of the PSD. The same discussion leads to the choice for the length of the source field (Fig. 3.10). The influence length of the source field is at maximum the diffraction length L_{total} in $+z$-direction. Therefore the minimum source length is chosen to be the diffraction length. This diffraction length is windowed in order for the wrap around effect to be negligible. The window function applied is a Tukey window [Har78]. The initial source field for the DSD is given as

$$u_{\text{Source}}^{(D)} = u_{\text{Source}}(\vec{r}) = A_{mn} H_m^{(1)}(k_\rho R_0) e^{-jm\varphi'} e^{-jk_z z'} \tag{3.25}$$

where $H_m^{(1)}(k_\rho R_0)$ is the Hankel function of first kind and order m. We now

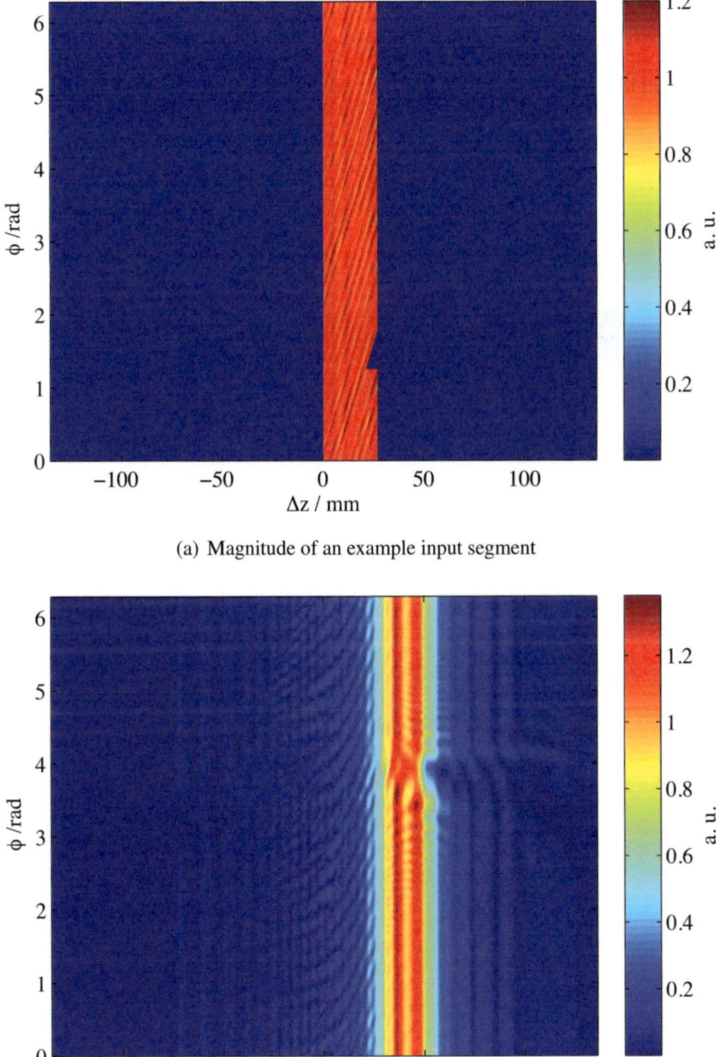

(a) Magnitude of an example input segment

(b) Magnitude of the resulting convolution of the example segment

Figure 3.17: Example convolution for one of the segments

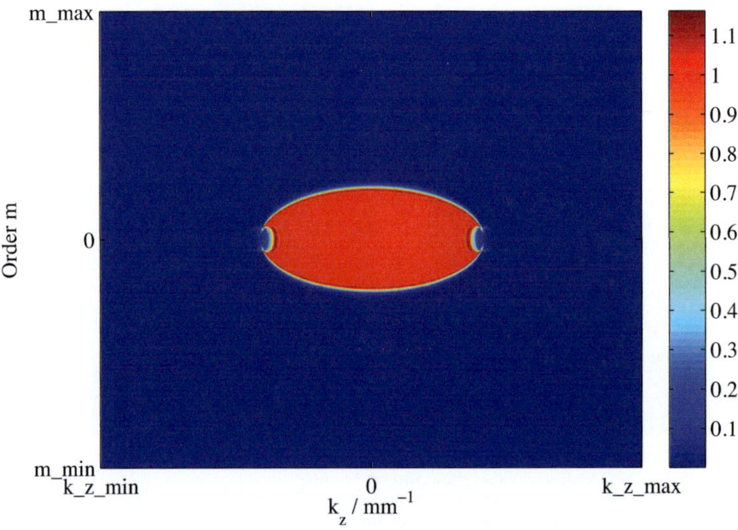

Figure 3.18: Magnitude of the FFT of an example DSD kernel

discuss the properties of the propagation by investigating the FFT of the kernel $g^{(D)}$, corresponding to the transfer function. The FFT of the kernel $g^{(D)}$ is given in magnitude and phase in figures 3.18 and 3.19 respectively. The form resembles the one from the PSD, except for a cut out piece near the edge of the transfer function. The parts of the spectrum at this far end can be neglected, since this corresponds to modes far beyond "cut off" and are not considered when employing the phase corrector model. One particular, but very important difference is the magnitude of the transfer function, the magnitude[9] has a value of $\left| \mathcal{F}_d\{g^{(D)}\} \right| \approx 1.0$. This corresponds to loss-free transmission of the field through the waveguide. Identical to the PSD, each spectral component of the field gets its individual phase shift (Fig. 3.19). So the propagation corresponds to a loss-free filtering of the field,

---

[9]The possibility of windowing the kernel, in order to reduce the ripple of the magnitude, is given in the implementation, but not discussed here.

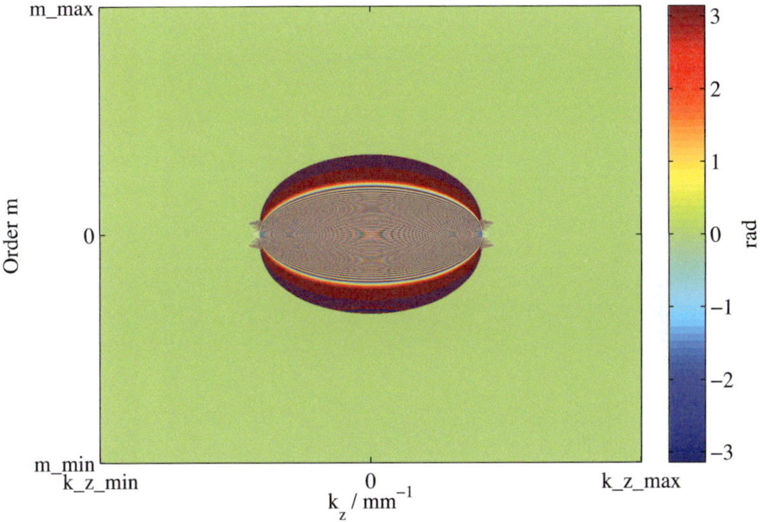

Figure 3.19: Phase of the FFT of an example DSD kernel

where each component exhibits an individual phase shift. The resulting field is again the superposition of all the filtered spectral components of the field. This concludes the discussion of the numerical aspects of the DSD. The simulation results will be discussed and compared in chapter 4. Now we come to the third method for comparison, the CWD.

3.4 Numerical aspects of the CWD

The CWD has also been implemented together with the DSD in the aforementioned C++ program. The iterative algorithm to solve for $\psi^{(1)}$ in equation (2.123), as well as the convolution algorithm are identical to the ones from the DSD, since both equations are iteratively solved using the Born approximation. Only the

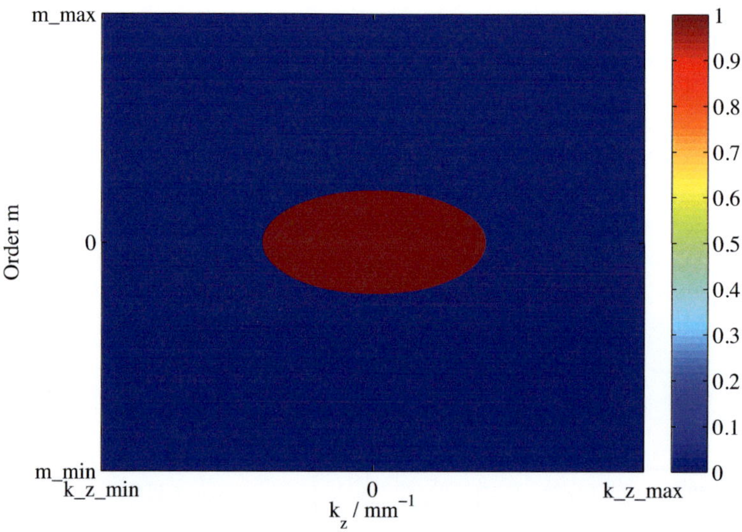

Figure 3.20: Amplitude of an example transfer function

integration kernel and therefore the transfer function changes. In case of the CWD, the transfer function is given explicitly without the need to perform an FFT. We will now discuss the transfer function of this method in order to describe the properties of the propagation.

Transfer function and plane wave decomposition

The similarities with system theory are evident after consideration of equations (2.124) and (2.125). Therefore we can consider $\mathcal{T}$ as the transfer function of the system "waveguide". Since we consider the propagation in an oversized waveguide with a rather high operating mode, we can simplify the spectrum of the transfer function by considering only modes with high eigenvalues and orders m

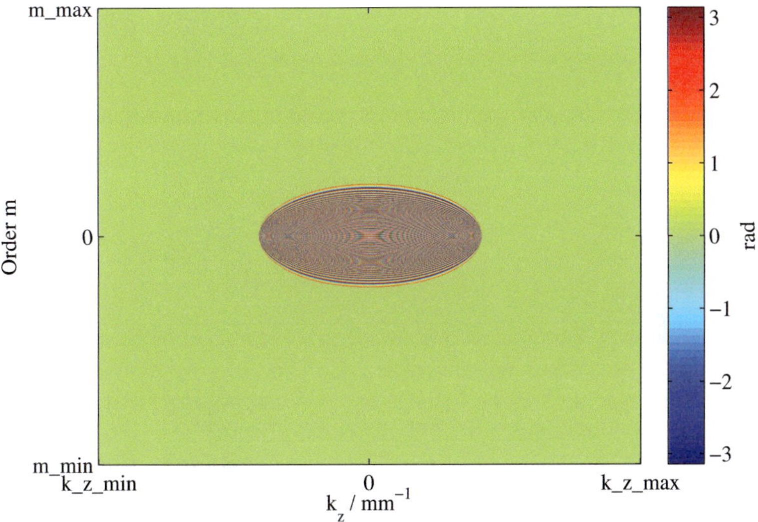

Figure 3.21: Phase of an example transfer function

that are smaller than the eigenvalue $(m < R_0\sqrt{k^2 - k_z^2}))$[10]. For these modes the Hankel functions and therefor the transfer function can be approximated by:

$$\mathcal{T} = \frac{H_m^{(2)}\left(R_0\sqrt{k^2 - k_z^2}\right)}{H_m^{(1)}\left(R_0\sqrt{k^2 - k_z^2}\right)} \tag{3.26}$$

$$\approx e^{-j2\cdot\left(\sqrt{R_0^2\cdot(k^2-k_z^2)-m^2}-m\arccos\left(\frac{m}{R_0\sqrt{k^2-k_z^2}}\right)-\frac{\pi}{4}\right)} \tag{3.27}$$

---

[10]This is actually a necessary condition for the scalar treatment to be valid, as mentioned before and not necessary valid for whispering gallery modes.

considering that

$$H_m^{(2)}\left(R_0\sqrt{k^2-k_z^2}\right) = \left(H_m^{(1)}\left(R_0\sqrt{k^2-k_z^2}\right)\right)^* \qquad (3.28)$$

where * denotes the complex conjugation. In order to have non attenuated propagation, that is real wave numbers, the following two expressions have to be fulfilled:

$$k^2-k_z^2 > 0 \quad , \quad 1-\left(\frac{m}{R_0\sqrt{k^2-k_z^2}}\right)^2 > 0 \qquad (3.29)$$

This means that such modes are above cut-off in the given geometry at the given frequency. Using these restrictions not all orders m and all wave-numbers k_z have to be taken into account and the transfer function can be simplified. From the square root of the right expression of (3.29), namely from:

$$\sqrt{1-\left(\frac{m}{R_0\sqrt{k^2-k_z^2}}\right)^2} \overset{!}{=} 0 \qquad (3.30)$$

an equation for an ellipse can be derived:

$$\frac{m^2}{(kR_0)^2} + \frac{k_z^2}{k^2} = 1 \qquad (3.31)$$

This equation describes the ellipse of the transfer function for the CWD. Since the cut off criteria is not stringent, ellipses for the PSD and DSD have a smoother transition in the transfer function to zero amplitude. The ellipse equation expresses that all modes with $m < kR_0$ and $k_z < k$ can propagate in the oversized waveguide.

Figures 3.20 and 3.21 show the magnitude and the phase of an example transfer function, which is a two dimensional filter, similar to the first two scalar methods. The source field for the CWD is equivalent to the one from the DSD. The calculation of the field can now be described as follows:

- Fourier Transform of $\psi^{(1)}$ on the waveguide wall gives the mode composition

- Each mode in the waveguide gets propagated by multiplying it with its corresponding phase shift, i.e. multiplication of the spectrum with the transfer function

- Inverse Fourier Transform gives $\psi^{(2)}$

- Multiplication of the phase corrector $e^{-j\vartheta(\varphi,z)}$ gives the new $\psi^{(1)}$.

A similar modeling of the propagation concerning resonators is addressed in [Wei69]. The described mechanism can be regarded as a CWD, hence the name. This decomposition is similar to a plane wave decomposition, which is equivalent to Rayleigh-Sommerfeld diffraction [She67]. This indicates the equivalence of the CWD and the DSD. As before, we compare the gained simulation results in chapter 4.

3.5 Adaptation for calculating tapered average radius launchers

In the previous sections, the launchers to be calculated had a constant average radius R_0, which is necessary for the FFT convolution to give valid results, especially for the PSD. We have seen in section 2.2, that launchers with harmonic perturbations incorporate a tapered average radius to suppress the possibilities for unwanted cavities which can lead to spurious oscillations. Due to the tapered average radius $R(z) = R_0 + \alpha z$ of the launcher, the wave numbers k_z depend on the z-coordinate, and the evaluation of the diffraction integral by means of the FFT over the whole calculation domain is not possible for the implementation of the PSD [JTP$^+$09]. The convolution algorithm used for the propagation of the three scalar methods can be interpreted as a linear space invariant filtering. On introducing a tapered average radius $R(z)$, the algorithm can be interpreted as linear space variant filtering with a dependence of the integration kernel on the z position. This issue is addressed by adapting an algorithm for a segmented convolution with a time variant impulse response to the diffraction problem with tapered average radius.

This algorithm was introduced by Rabiner and Schafer for the processing of speech signals [Rab78]. It is also known as weighted overlap-add convolution algorithm [KK06] and has the following main aspects compared to the regular overlap-add algorithm:

- The input segments overlap and are windowed

- The overlap and the window function have to fulfill the constant overlap add criteria [Smi11]

- Each windowed input segment is convolved with the impulse response, corresponding to the input segment

- The results of the convolutions are summed according to their position.

The effect of the algorithm is smearing the change in the impulse response, from one segment to the next over the convolution domain. The differences to the regular segmented convolution are the existence of more than one impulse response and the overlapping of the input segments.

By adapting this algorithm to the segmented convolution algorithms for the DSD and the CWD method, we are able to calculate launchers with tapered average radius. In this work, only the adaptation of the algorithm for the DSD is presented, since the application of the algorithm to the CWD is straight forward. The details of the adaptation consist of:

Hann windowing of the input segments: The segments used for the tapered average radius have a length of two Brillouin length L_B without the zero padding in order to allow an overlap of one Brillouin length. Therefore the Hann window applied has a length of two L_B. An example of the used Hann[11] window [Har78] is shown in figure 3.22. The overlap is best explained in one dimension only and depicted in figure 3.23. The window function is shifted from one segment to the next by one Brillouin length for the segments to overlap by one Brillouin length L_B.

Convolve each segment with its individual Green's function: Each segment is than convolved with its corresponding Green's function. Equation 3.21 gives the individual Green's function with the corresponding radius for each segment. Table 3.2 gives the radii for the tapered 118 GHz launcher example in section 2.2.1. Figure 3.24 gives the magnitude of a windowed example input segment and figure 3.25 gives the magnitude of the resulting convolution of this segment. In

---

[11] named after Julius von Hann, also known as Hanning window

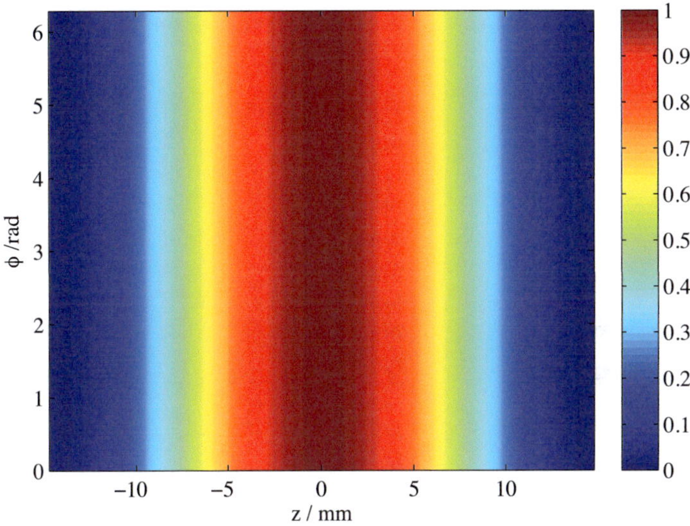

Figure 3.22: Example Hann window for the weighted overlap add algorithm

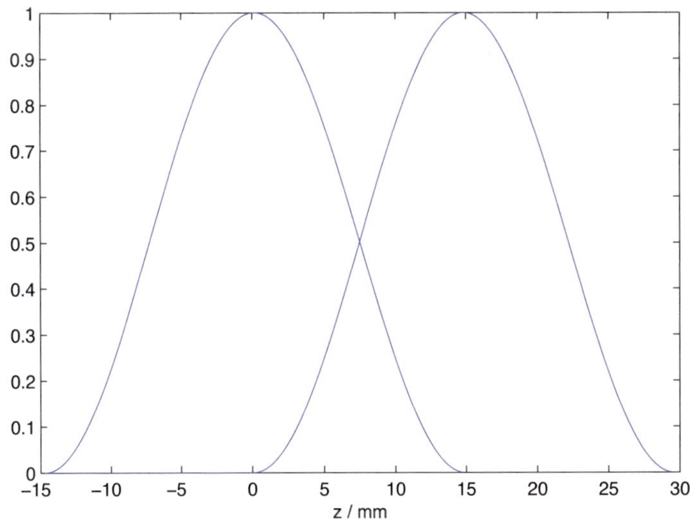

Figure 3.23: Overlapping of the window functions of two consecutive input
          segments

this example the propagation from one Brillouin zone to the next in one iteration is illustrated.

Segment	$R(z)$	Segment	$R(z)$
0	20.03 mm	8	20.26 mm
1	20.06 mm	9	20.29 mm
2	20.09 mm	10	20.32 mm
3	20.12 mm	11	20.35 mm
4	20.15 mm	12	20.38 mm
5	20.18 mm	13	20.41 mm
6	20.21 mm	14	20.44 mm
7	20.23 mm		

Table 3.2: Radius $R(z)$ for each segment of the tapered $TE_{22,6}$-mode launcher from section 2.2.1

Application of the taper to each segment after the convolution: Each convolved segment is multiplied with the phase correction corresponding to the taper in each segment. This accounts for the conical opening of the waveguide. The phase of an example phase correction for the taper is shown in figure 3.26. The z position of the phase correction is relative to the input segment. Identically to the un-tapered algorithm the phase corrector due to the perturbation is applied to give the new ingoing wave. In summary, the algorithm consists of three adaptations to the original segmented overlap-add convolution:

1. Windowing of the input segments with a window function fulfilling the COLA criteria.

2. Convolve each segment with its individual Green's function

3. Application of the taper to each segment after the convolution to account for the conical shape

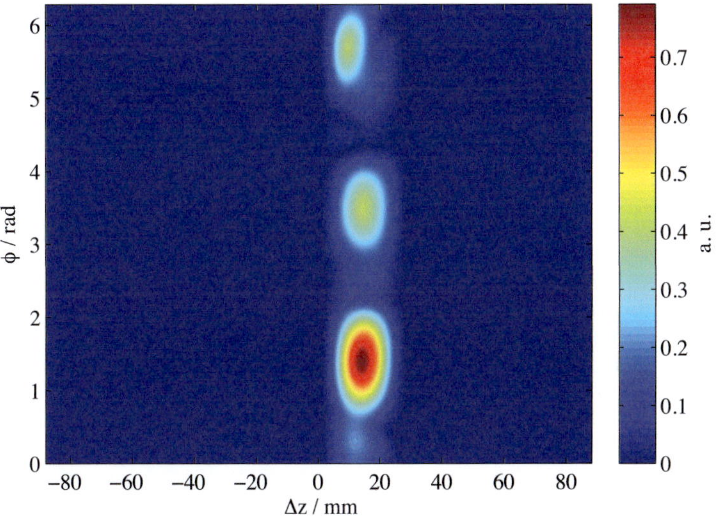

Figure 3.24: Magnitude of a windowed example input segment

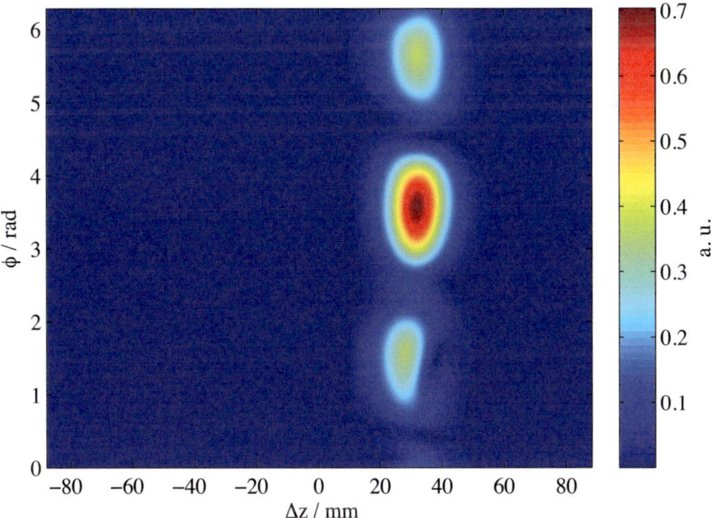

Figure 3.25: Magnitude of the resulting convolution of the windowed example
segment

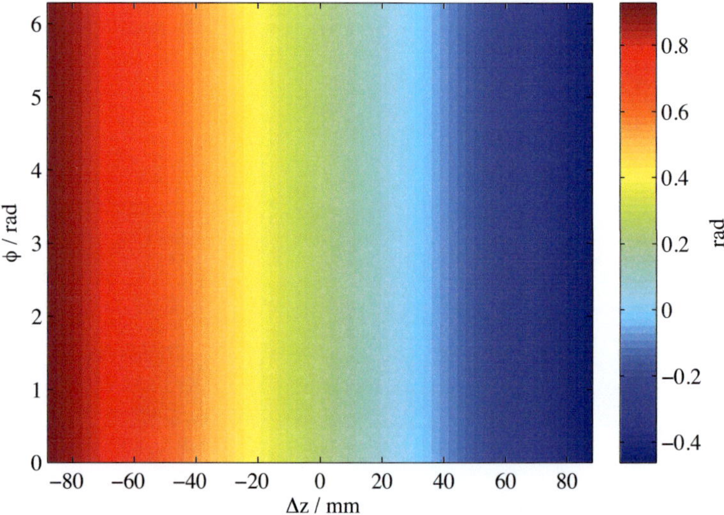

Figure 3.26: Example phase of the phase correction due to the conical opening of
the waveguide

3.6 Numerical aspects of the EFIE

The solver for the EFIE is implemented in a commercial computer program named
Surf3D [Nei04]. The computer program solves the EFIE using the Methods
of Moments (MoM) [Har68, PRM98]. Since the electromagnetic problems are
solved in all three dimensions, they are very large[12]. Consequently, the necessary
matrix-vector multiplication for the iterative solution is sped up by employing
the Multilevel Fast Multipole Algorithm (MLFMA) to be able to calculate these
problems on a workstation computer. We now describe the two used numerical
methods, MoM and MLFMA, starting with the MoM.

---

[12]typical memory size for a launcher calculation is over 7 GB, for the whole quasi-optical converter
up to 50 GB

3.6.1 Method of Moments

In order to solve equation (2.143) for the unknown current density $\vec{J}_S$, we rewrite the EFIE into an operator form, similar to equations (2.113) and (2.114):

$$\mathcal{L}f = g \tag{3.32}$$

where $f \equiv \vec{J}_S$ denotes the unknown current density, $g \equiv \vec{E}_t^I$ the tangent incident electric field and $\mathcal{L}$ denotes the operator including the differentiations, integrations and summation. Next, we approximate f by a series solution as [PRM98]:

$$f \cong \sum_{n=1}^{N} \alpha_n B_n \tag{3.33}$$

where B_n are the chosen basis functions and α_n denote the unknown coefficients. When using δ-functions as basis function, the method is referred to as Point-Matching [Bal89]. Substituting (3.33) into (3.32) gives:

$$\sum_{n=1}^{N} \alpha_n \mathcal{L}B_n = g \tag{3.34}$$

Now forcing the residual of the projection of (3.34) onto a set of testing functions T_m, to be zero, we obtain:

$$\sum_{n=1}^{N} \alpha_n \langle T_m, \mathcal{L}B_n \rangle - \langle T_m, g \rangle = 0 \quad m = 1, 2, \ldots, N \tag{3.35}$$

where the projection of two functions w and g onto a domain L corresponds to:

$$\langle w, g \rangle = \iint_L w^* \cdot g \, dL \tag{3.36}$$

Now letting:

$$l_{mn} = \langle T_m, \mathcal{L}B_n \rangle \tag{3.37}$$

$$\beta_m = \langle T_m, g \rangle \tag{3.38}$$

we finally obtain the following matrix equation of dimension $N \times N$:

$$\sum_{n=1}^{N} l_{mn}\alpha_n = \beta_m \tag{3.39}$$

Solving this matrix equation by inversion

$$\alpha_n = \sum_{m=1}^{N} l_{mn}^{-1}\beta_m \tag{3.40}$$

gives the coefficients α_n for $f \equiv \vec{J_S}$ in equation (3.33). Together with the chosen basis function, we have determined the surface current density $\vec{J_S}$ and can calculate the scattered fields $\vec{E}^S$ and $\vec{H}^S$.

3.6.2 Multilevel Fast Multipole Method

Since the wavelength is much smaller than the dimensions of the geometry, the resulting size of the matrix to be inverted is very large. A direct inversion of equation (3.40) requires $\mathcal{O}(N^3)$ operations and $\mathcal{O}(N^2)$ amount of memory and is usually not practical for such large systems. An alternative solution is an iterative solution, e.g. a conjugate gradient algorithm, of equation (3.39) [PRM98]. This iterative solution requires an evaluation of the left hand side of (3.39) using matrix-vector multiplication. A fast way to implement this calculation is the Fast Multipole Method.

In this method, the scatterer, on which the evaluation takes place, is divided into N cells, that is N basis functions. These cells are then grouped into p segments. Now the interaction between segments can be classified into two zones. These zones are the near-zone and the far-zone and depend on the distance between the two segments p_i and p_j to be evaluated. In the near-zone the matrix-vector multiplication is done by direct evaluation of

$$l_{mn}\alpha_n = \langle T_m, \mathcal{L}B_n \rangle \alpha_n \tag{3.41}$$

where as in the far-zone a far-field approximation of the operator $\mathcal{L}$ is used. This approximation gives the possibility for a factorization of the matrix elements

$$l_{mn} = \gamma_{L'} \cdot l_{diag} \cdot \gamma_L \,, \tag{3.42}$$

where l_{diag} is a diagonal matrix and $\gamma_{L'}$ and γ_L are vectors. The factors in equation (3.42) can be interpreted as follows:

- $\gamma_{L'}$ corresponds to the contribution of the cell N_m to the far field radiation pattern of the segment p_i

- l_{diag} corresponds to the translation of the far field radiation pattern from p_i to the incoming pattern at p_j

- γ_L corresponds to the weighting of the incoming pattern at p_j contributing to the cell N_n.

This factorization reduces the computational complexity to $\mathcal{O}(N^{1.5})$ and can be further reduced to a cost of $\mathcal{O}(N \log_2 N)$ for both computation and memory in the case of sparse matrices [PRM98]. The factorization is also regarded as a multipole expansion, hence the name Fast Multipole Method (FMM). By nesting this algorithm into a hierarchical tree, one can extend the FMM to a multi level algorithm known as the Multilevel Fast Multipole Algorithm (MLFMA). For an in depth explanation of the FMM and MLFMA, see [Che01].

4 Simulation results and comparison

In this chapter the verification of the two newly implemented methods, the DSD and the CWD, is performed by comparison of the calculation results of an existing launcher design with the commercial full wave code Surf3D. Furthermore, a comparison with the existing scalar diffraction code, TWLDO, implementing the PSD, is performed. TWLDO was used to synthesize the launcher surface employed for the verification and was kindly provided by Jin [Jin10]. The results are presented in following order: The first result is presented for the PSD, followed by the DSD result. Subsequently the result for the CWD is given and finally the result for the EFIE is presented. The criteria for comparing the fields are the correlation coefficients introduced in chapter 2 in (2.79) and (2.80). Besides the correlation of the different results, each result is correlated to the Gaussian target function used in the synthesis.

The second part of this chapter is dedicated to the calculation results of the tapered average radius launchers. In order to verify the algorithm, the DSD is compared to the coupled mode equations method for the case of the periodic wall perturbation, since the PSD implementation does not give the possibility for tapered geometries. A further example for an adaptive surface perturbation is given, in which the DSD is compared to the EFIE. The CWD is not implemented for tapered geometries due to the high similarity to the DSD.

4.1 Launcher with constant average radius

The main parameters of the launcher used in the comparison are the same as for the example launcher used in section 3.3. For easier reference, they are repeated in table 4.1. Figure 4.1 shows the geometry of the launcher with a magnification of the perturbation by a factor of 10, without the magnification the perturbation is barely visible. This gives a good idea of the complexity of the surface. The unrolled inner launcher surface $\Delta R(\varphi, z)$ used in all the simulations is given in

Parameter	Value
Operating mode	$TE_{34,19}$
Frequency f	170 GHz
Launcher radius R_0	32.5 mm
Taper angle $\tan\alpha$	0.000
Launcher length L	$z = 291.6$ mm
Beginning of cut $L_{Start} = L - L_C$	$z = 220.0$ mm

Table 4.1: Parameters of the launcher for the comparison

figure 4.2. From the structure of the perturbation, the curved mirror like surface for each Brillouin zone can also be identified very well. For the three scalar models, the iterative solutions are presented. For the EFIE the final result is presented, since no intermediate results are available. The magnitude of the Gaussian target function used for the complex correlation is depicted in figure 4.3.

4.1.1 PSD result

Performing the calculation procedure for the case of the PSD according to (2.111) gives the iterative results $u_i^{(P)}$. The magnitudes of these results are shown in figures 4.4 and 4.5. The initial field, on the whole launcher wall, is the given cavity mode. For the first iteration in figure 4.4(a), one can identify a Fresnel type field in all Brillouin zones across the launcher wall. This type of field is expected from an initial constant field magnitude. The reduction of the field by a factor of 2, as indicated in the discussion of the transfer function, is reflected in the magnitudes of each iteration step. As the iteration steps increase, the field turns into the desired field distribution. The final calculation result $u_i^{(P)}$ from the last (13-th) iteration is shown in figures 4.6 and 4.7 in magnitude and phase, respectively. The result shows a good field bunching on the launcher wall. From figure 4.6, the focusing and forming of the Brillouin zones can be observed. The determination of the complex correlation coefficient (2.80) to the Gaussian target function is performed over the radiating Brillouin zone depicted by the red box in figure 4.6. Additionally figure 4.6, shows the well focused field in the radiating

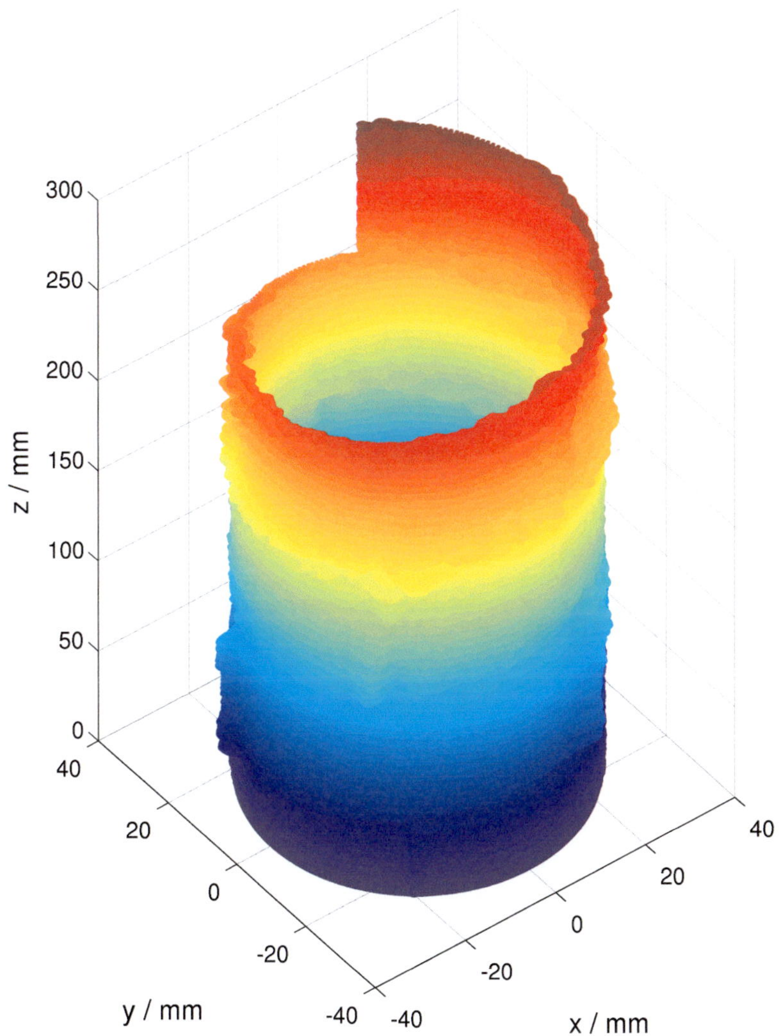

Figure 4.1: Geometry of the $TE_{34,19}$-mode 170 GHz launcher used for the comparison (perturbation 10x magnified)

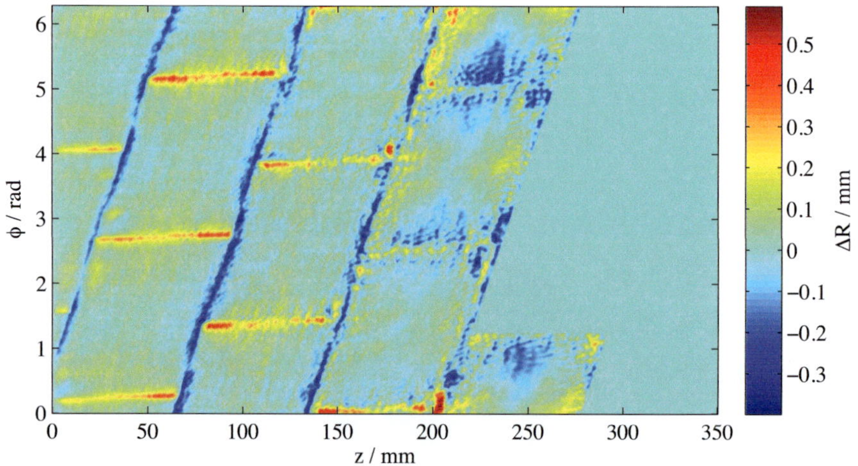

Figure 4.2: Perturbation of the inner launcher surface

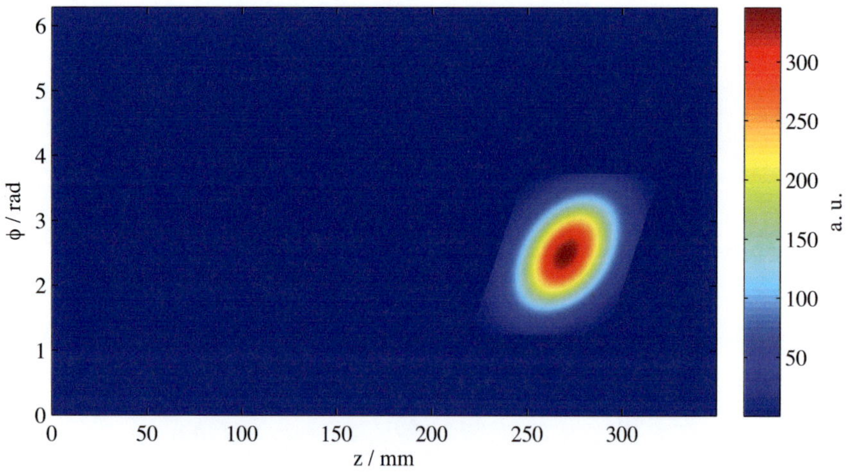

Figure 4.3: Magnitude of the Gaussian target function used for the correlation

Parameter	Value
Matrix size in φ-direction	1024
Matrix size in z-direction	4096
Size of calculation domain [mm]	$0 \leq z \leq 350.15$
Size of source + calculation domain [mm]	$-1050.45 \leq z \leq 350.15$
Sampling in φ-direction [rad]	$\Delta\varphi = 0.006$
Sampling in z-direction [mm]	$\Delta z = 0.342$
Vector correlation to Gaussian target function	$\eta_{vector} = 98.3\%$
Scalar correlation to Gaussian target function	$\eta_{scalar} = 98.7\%$
Iterations for final result	13
Time for computation	5.5 minutes

Table 4.2: Main calculation parameters for the PSD

aperture (Brillouin zone). The parameters of the calculation are summarized in table 4.2.

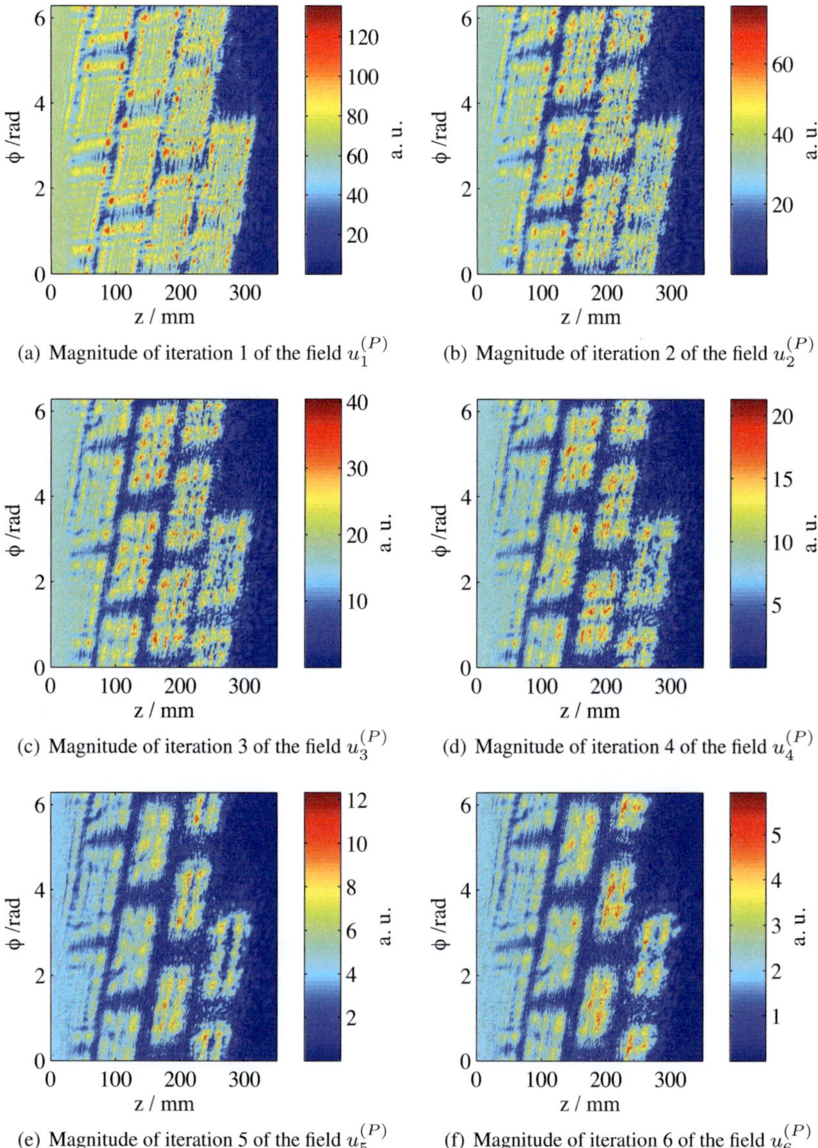

(a) Magnitude of iteration 1 of the field $u_1^{(P)}$

(b) Magnitude of iteration 2 of the field $u_2^{(P)}$

(c) Magnitude of iteration 3 of the field $u_3^{(P)}$

(d) Magnitude of iteration 4 of the field $u_4^{(P)}$

(e) Magnitude of iteration 5 of the field $u_5^{(P)}$

(f) Magnitude of iteration 6 of the field $u_6^{(P)}$

Figure 4.4: Magnitude of $u^{(P)}$ for the iterations 1-6

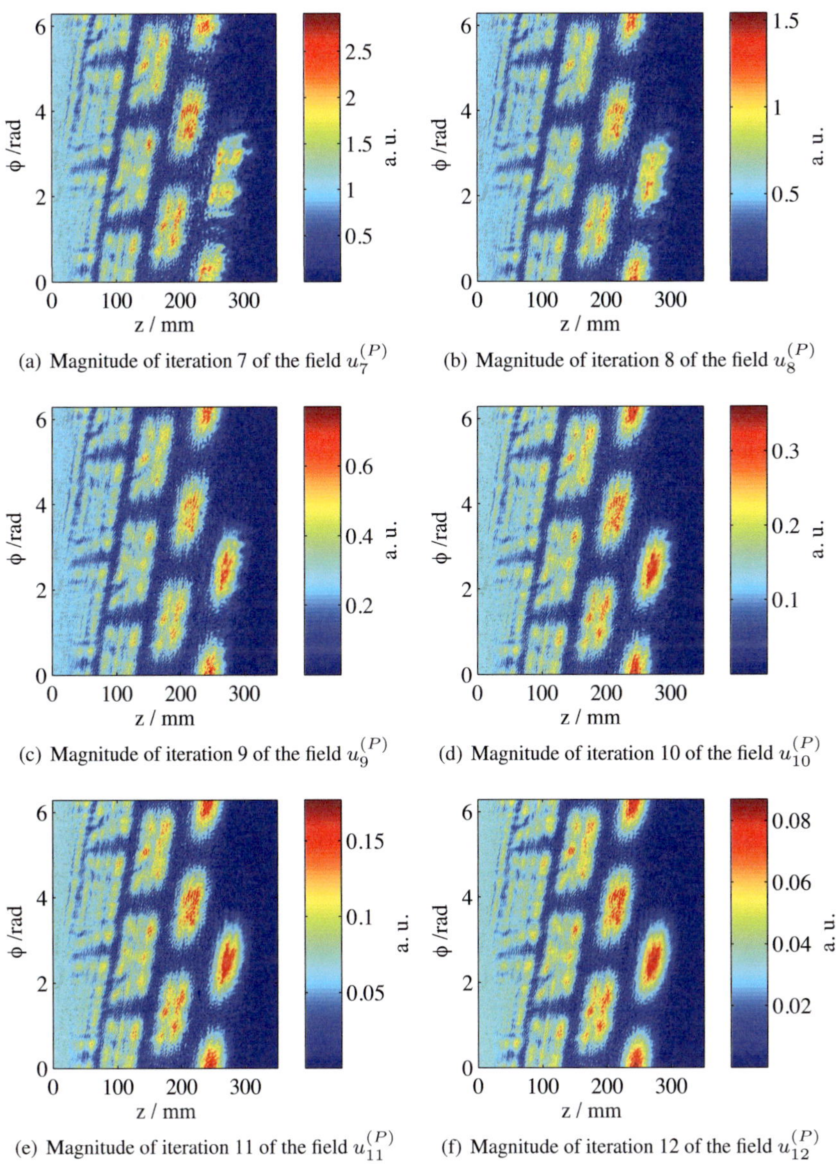

(a) Magnitude of iteration 7 of the field $u_7^{(P)}$

(b) Magnitude of iteration 8 of the field $u_8^{(P)}$

(c) Magnitude of iteration 9 of the field $u_9^{(P)}$

(d) Magnitude of iteration 10 of the field $u_{10}^{(P)}$

(e) Magnitude of iteration 11 of the field $u_{11}^{(P)}$

(f) Magnitude of iteration 12 of the field $u_{12}^{(P)}$

Figure 4.5: Magnitude of $u^{(P)}$ for iterations 7-12

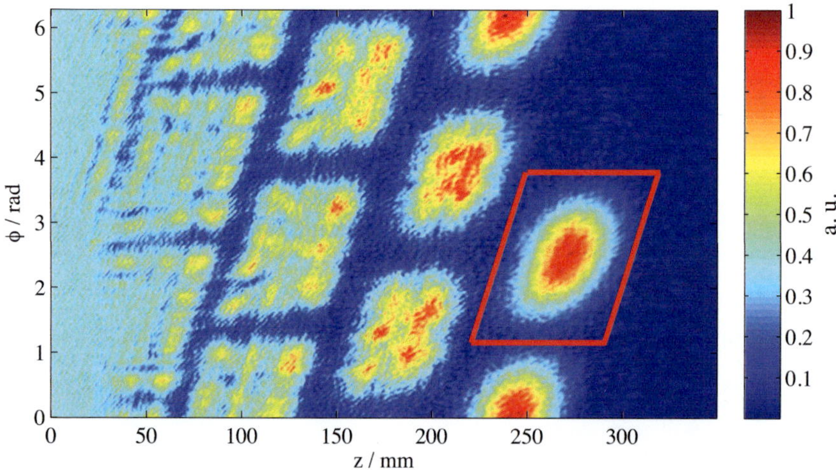

Figure 4.6: Magnitude (normalized) of the field $u_{final}^{(P)} \equiv u_{13}^{(P)}$ resulting from the perturbed wall

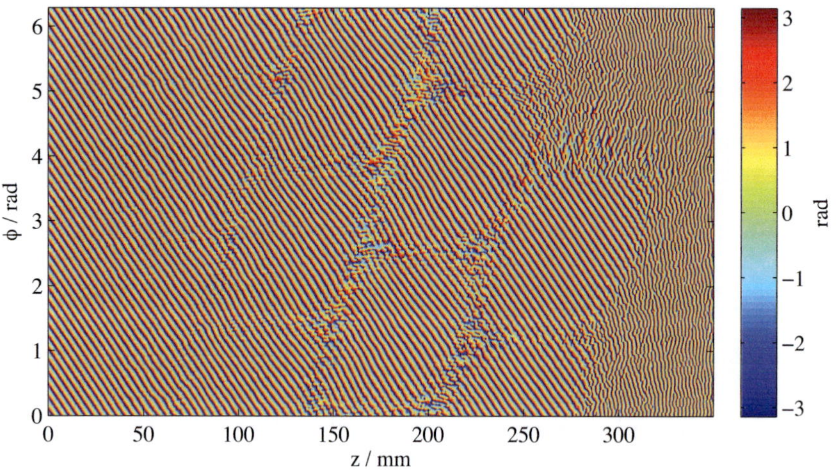

Figure 4.7: Phase of the field $u_{final}^{(P)} \equiv u_{13}^{(P)}$ resulting from the perturbed wall

4.1.2 DSD result

For the DSD, we present the iterative solution[1] of $u_i^{(D)}$, resulting from the algorithm explained in chapter 3. The intermediate and the final solutions show the outward traveling field, before the application of the phase corrector, i. e. the presented results are the magnitude of the field $u_i^{(D)}$ after the evaluation of the integral from equation 2.112. Due to the amount of iterations, only the magnitude for the intermediate results are depicted. Additionally, we neglect iterations 18 and 19 for space reasons. Figures 4.8(a) through 4.10(f) and the final result (Figs. 4.11 and 4.12) illustrate the wave propagation from the beginning of the launcher towards the end. The formation of the field distribution along the waveguide wall can be observed. The higher the iteration number, the further the field has propagated. The final calculation result (after 20 iterations) is given in both magnitude and phase. As for the PSD, the desired field bunching due to the perturbation occurs. Comparing figures 4.11 and 4.6 slight differences in the magnitudes can be identified. This results from one of the main differences in the implementation between the PSD and the DSD. This difference is the initial field. For the PSD, the main cavity mode is given as initial field distribution on the whole launcher wall. This is different for the DSD. Here the incident wave from the cavity into the launcher region is taken as the initial field distribution (Fig. 4.8(a)). The parameters for the calculation of the DSD are summarized in table 4.3. Since we use a segmented convolution algorithm, the number of parameters differs from the PSD. The biggest difference is the resulting computational time, which is reduced by a factor of more than 5, even though more iterations are used. Table 4.3 shows 60 seconds for the DSD. The PSD uses only 13 iterations for the final solution. For this amount of iterations, the DSD algorithm has only a computational time of 48 seconds.

---

[1]Our choice of starting from iteration number 0 is due to Born's formulation of the 0-th iteration being the field incident into the scattering region.

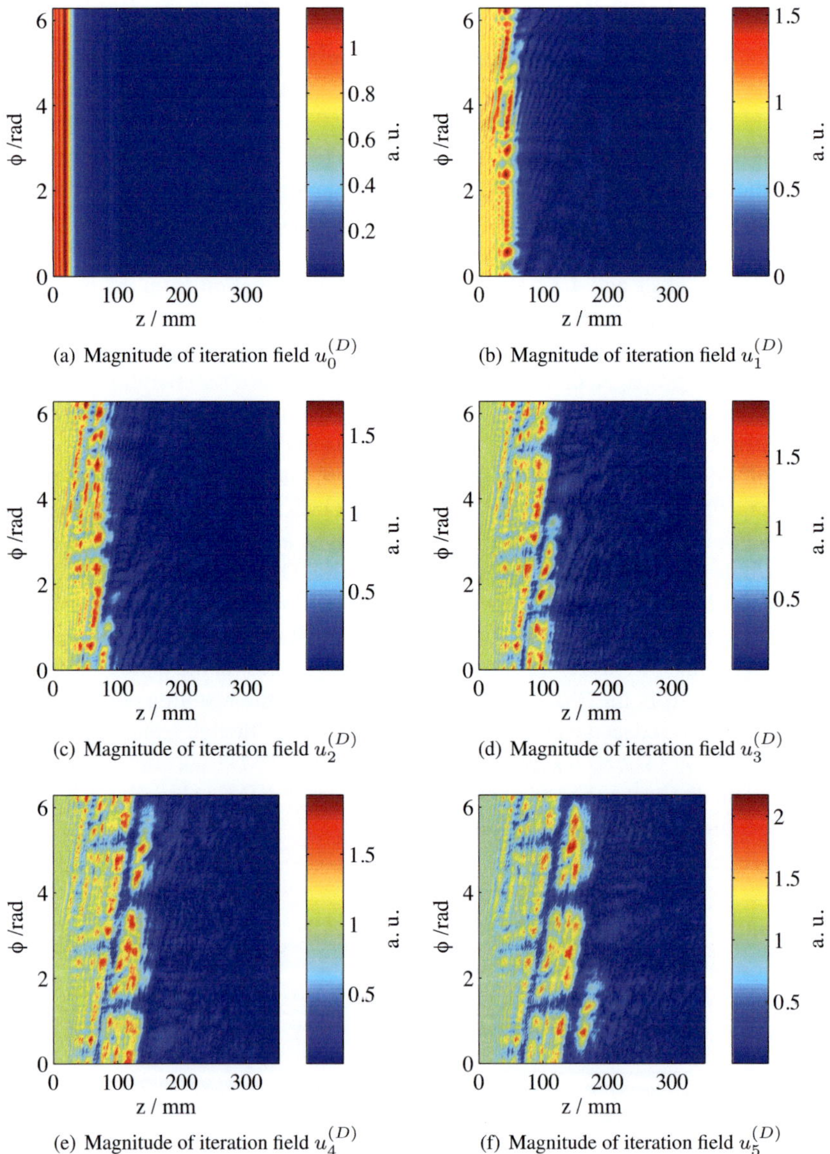

(a) Magnitude of iteration field $u_0^{(D)}$

(b) Magnitude of iteration field $u_1^{(D)}$

(c) Magnitude of iteration field $u_2^{(D)}$

(d) Magnitude of iteration field $u_3^{(D)}$

(e) Magnitude of iteration field $u_4^{(D)}$

(f) Magnitude of iteration field $u_5^{(D)}$

Figure 4.8: Magnitude of $u^{(D)}$ for iterations 0-5

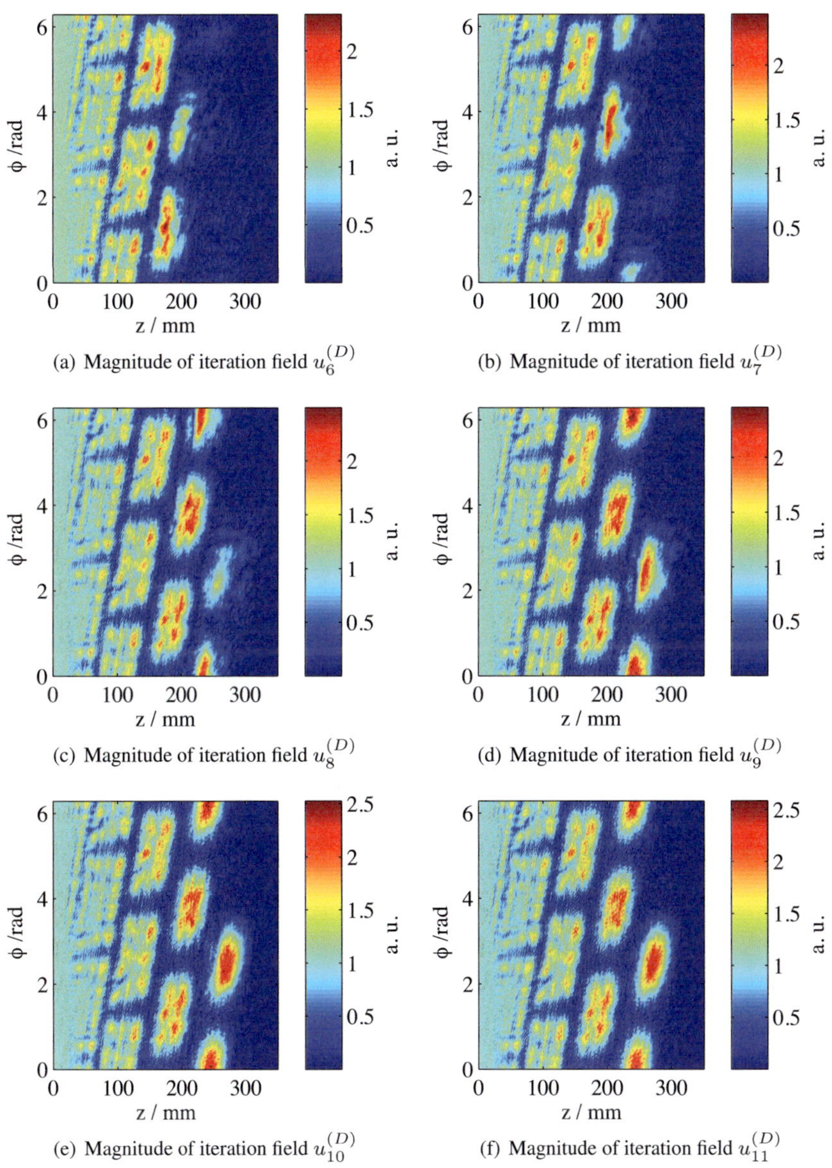

(a) Magnitude of iteration field $u_6^{(D)}$

(b) Magnitude of iteration field $u_7^{(D)}$

(c) Magnitude of iteration field $u_8^{(D)}$

(d) Magnitude of iteration field $u_9^{(D)}$

(e) Magnitude of iteration field $u_{10}^{(D)}$

(f) Magnitude of iteration field $u_{11}^{(D)}$

Figure 4.9: Magnitude of $u^{(D)}$ for iterations 6-11

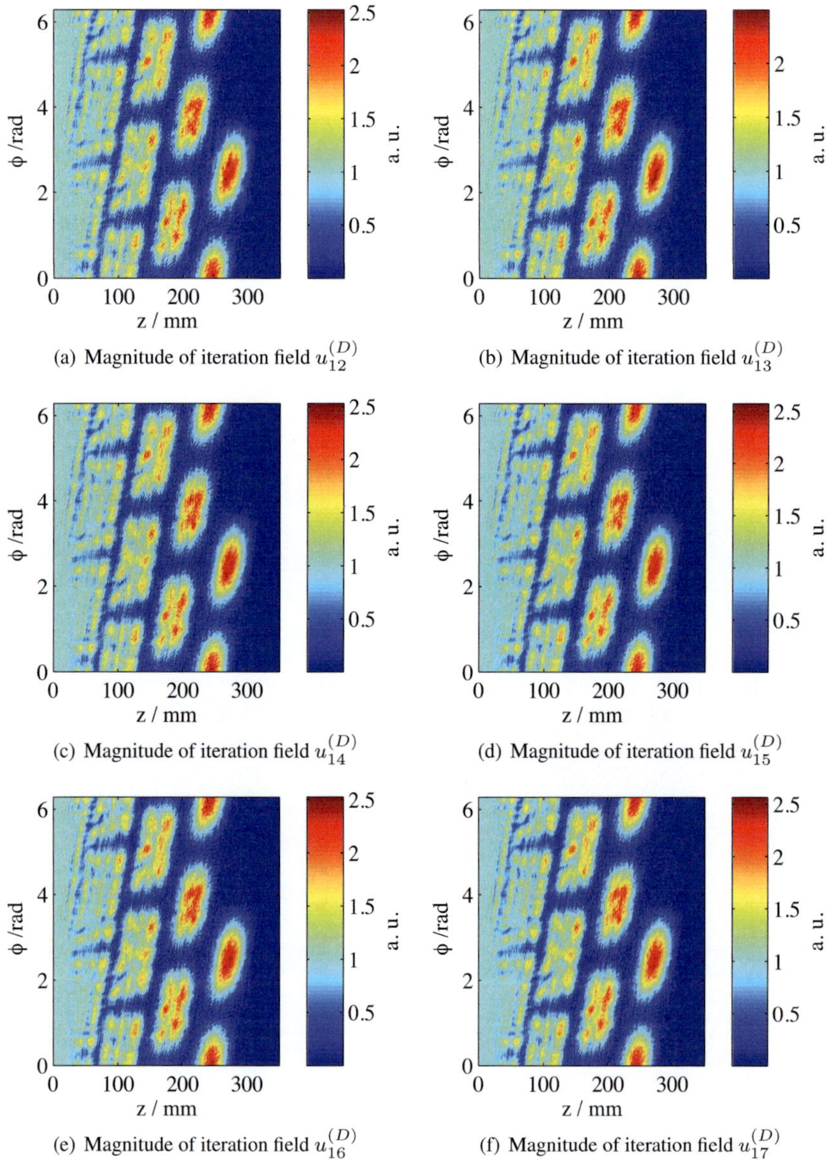

(a) Magnitude of iteration field $u_{12}^{(D)}$

(b) Magnitude of iteration field $u_{13}^{(D)}$

(c) Magnitude of iteration field $u_{14}^{(D)}$

(d) Magnitude of iteration field $u_{15}^{(D)}$

(e) Magnitude of iteration field $u_{16}^{(D)}$

(f) Magnitude of iteration field $u_{17}^{(D)}$

Figure 4.10: Magnitude of $u^{(D)}$ for iterations 12-17

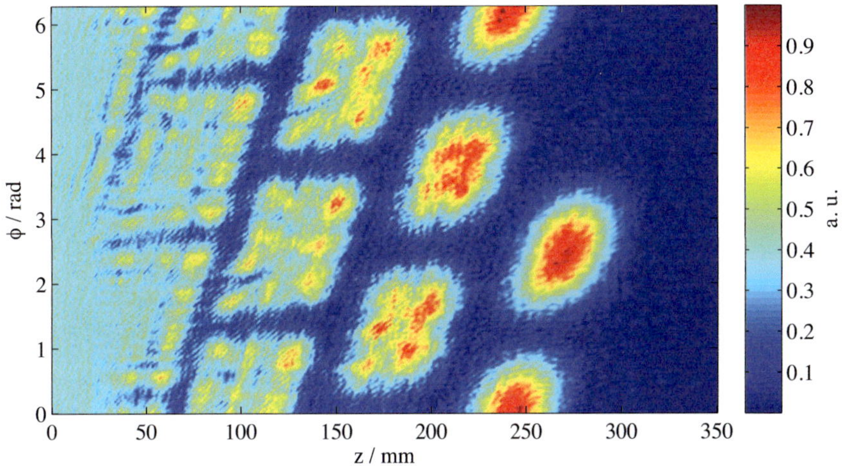

Figure 4.11: Magnitude (normalized) of the field $u_{final}^{(D)} \equiv u_{20}^{(D)}$ resulting from the perturbed wall

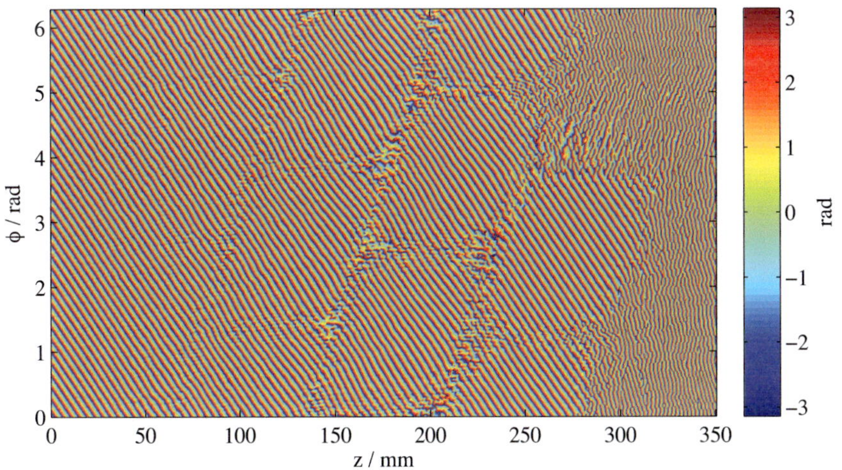

Figure 4.12: Phase of the field $u_{final}^{(D)} \equiv u_{20}^{(D)}$ resulting from the perturbed wall

Parameter	Value
For the launcher domain:	
Matrix size in φ-direction	1024
Matrix size in z-direction	1024
Length of launcher domain [mm]	$0 \leq z \leq 350.15$
For initial convolution:	
Source matrix size in z-direction	1273
Length of source + calculation domain [mm]	$-84.91 \leq z \leq 350.15$
For FFT domain:	
Size of FFT in z-direction due to segmentation	830
Size of FFT in φ-direction	1024
Sampling in φ-direction [rad]	$\Delta\varphi = 0.006$
Sampling in z-direction [mm]	$\Delta z = 0.342$
No. of iterations for the final result	20
Vector correlation to Gaussian target function	$\eta_{vector} = 98.2\%$
Scalar correlation to Gaussian target function	$\eta_{scalar} = 98.7\%$
Time for computation	60 seconds

Table 4.3: Main calculation parameters for the DSD

4.1.3 CWD result

For the third approach, we use the corresponding method described in sections
2.3.4 and 3.4 for the calculation, and show the successively iterated solution for
the given launcher with the wall deformation from figure 4.2. Figure 4.13(a)
shows the field incident into the launcher. The following figures 4.13(b) through
4.15(f) show the iterations of $\psi_i^{(2)}$ corresponding to the field propagation through
the waveguide. Again, due to space reasons, we neglect iterations 18 and 19.
Figures 4.16 and 4.17 show the final result in magnitude and phase. The results
of the CWD look very similar to the results of the DSD and the PSD with slight
differences (Figs. 4.16, 4.11 and 4.6) . The number of iterations used for the
calculation is the same as for the DSD. More details on the comparison are given
in section 4.1.5. The main calculation parameters for the CWD are summarized in
table 4.4 and are identical to the DSD except for the transfer function used in the
calculation.

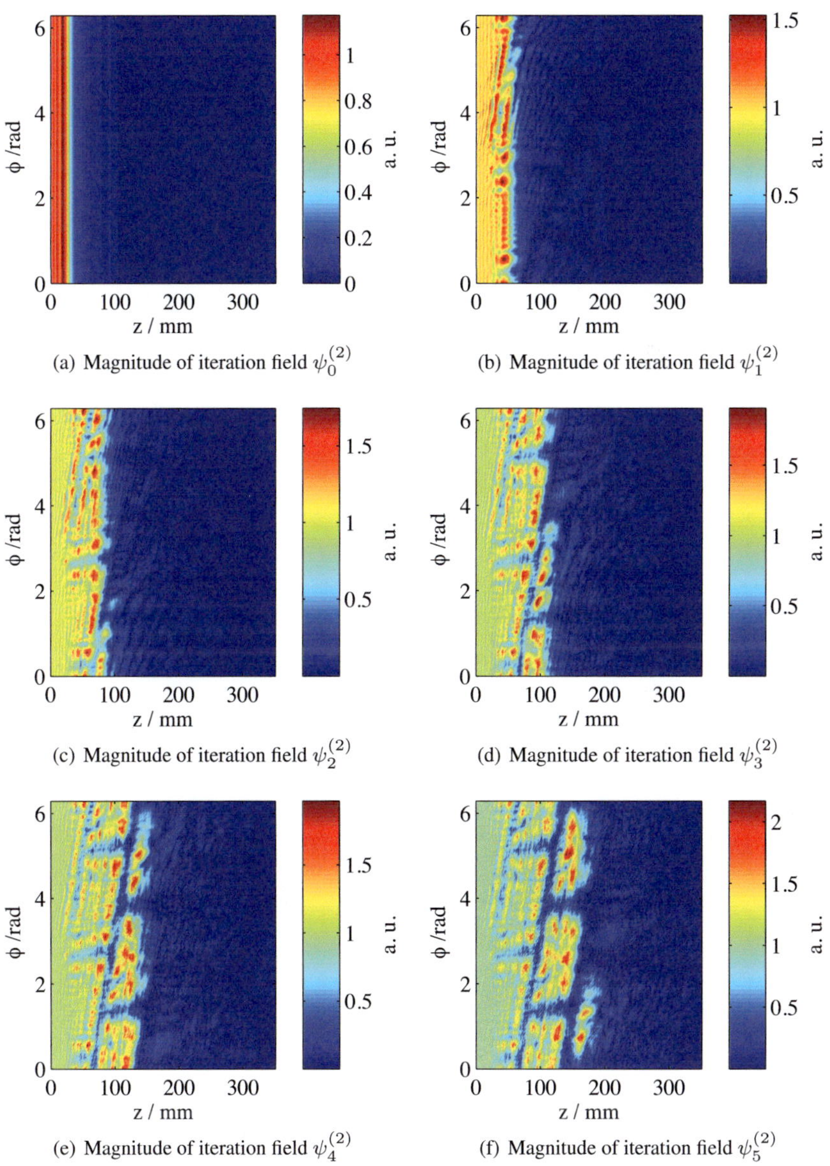

(a) Magnitude of iteration field $\psi_0^{(2)}$

(b) Magnitude of iteration field $\psi_1^{(2)}$

(c) Magnitude of iteration field $\psi_2^{(2)}$

(d) Magnitude of iteration field $\psi_3^{(2)}$

(e) Magnitude of iteration field $\psi_4^{(2)}$

(f) Magnitude of iteration field $\psi_5^{(2)}$

Figure 4.13: Magnitude of $\psi^{(2)}$ for iterations 0-5

107

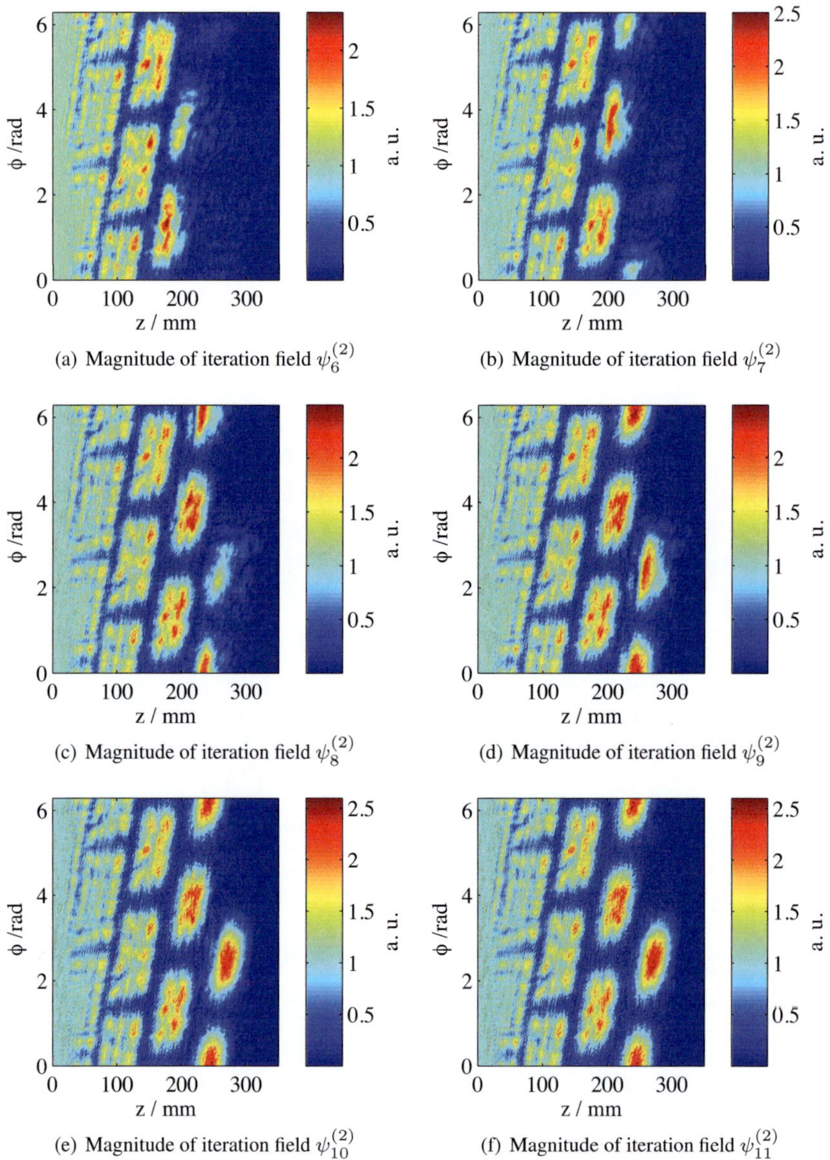

(a) Magnitude of iteration field $\psi_6^{(2)}$

(b) Magnitude of iteration field $\psi_7^{(2)}$

(c) Magnitude of iteration field $\psi_8^{(2)}$

(d) Magnitude of iteration field $\psi_9^{(2)}$

(e) Magnitude of iteration field $\psi_{10}^{(2)}$

(f) Magnitude of iteration field $\psi_{11}^{(2)}$

Figure 4.14: Magnitude of $\psi^{(2)}$ for iterations 6-11

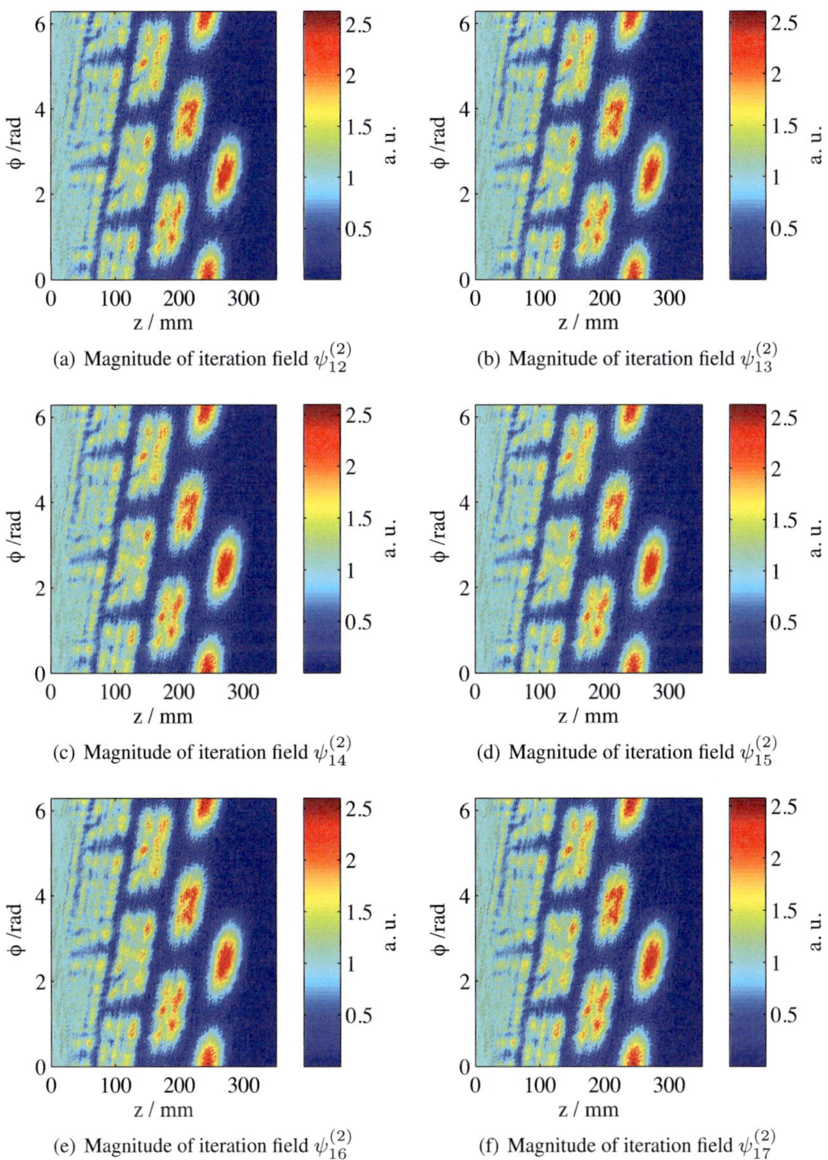

(a) Magnitude of iteration field $\psi_{12}^{(2)}$

(b) Magnitude of iteration field $\psi_{13}^{(2)}$

(c) Magnitude of iteration field $\psi_{14}^{(2)}$

(d) Magnitude of iteration field $\psi_{15}^{(2)}$

(e) Magnitude of iteration field $\psi_{16}^{(2)}$

(f) Magnitude of iteration field $\psi_{17}^{(2)}$

Figure 4.15: Magnitude of $\psi^{(2)}$ for iterations 12-17

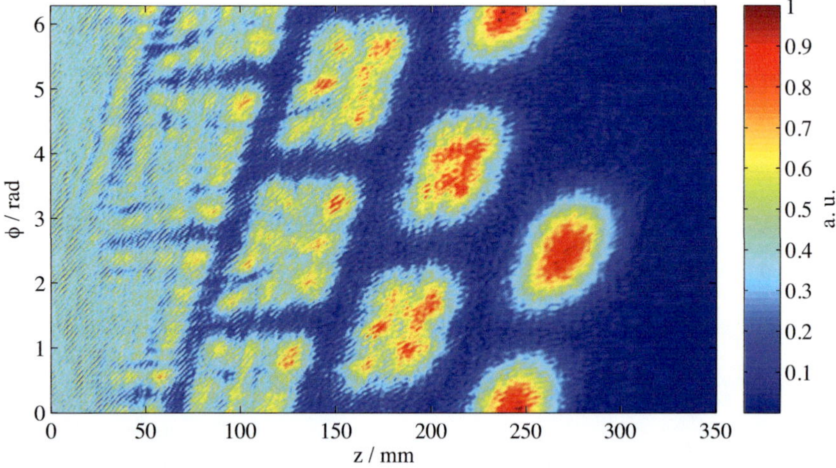

Figure 4.16: Magnitude (normalized) of the field $\psi_{final}^{(2)} \equiv \psi_{20}^{(2)}$ resulting from the perturbed wall

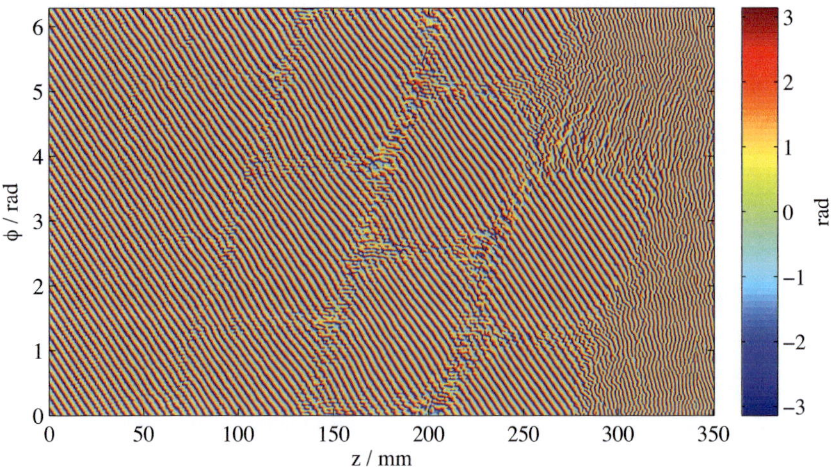

Figure 4.17: Phase of the field $\psi_{final}^{(2)} \equiv \psi_{20}^{(2)}$ resulting from the perturbed wall

Parameter	Value
For the launcher domain:	
Matrix size in φ-direction	1024
Matrix size in z-direction	1024
Length of launcher domain [mm]	$0 \leq z \leq 350.15$
For initial convolution:	
Source matrix size in z-direction	1273
Length of source + calculation domain [mm]	$-84.91 \leq z \leq 350.15$
For FFT domain:	
Size of FFT in z-direction due to segmentation	830
Size of FFT in φ-direction	1024
Sampling in φ-direction [rad]	$\Delta\varphi = 0.006$
Sampling in z-direction [mm]	$\Delta z = 0.342$
No. of iterations for the final result	20
Vector correlation to Gaussian target function	$\eta_{vector} = 98.0\%$
Scalar correlation to Gaussian target function	$\eta_{scalar} = 98.7\%$
Time for computation	60 seconds

Table 4.4: Main calculation parameters for the CWD

4.1.4 EFIE result

The solution resulting from the EFIE, as calculated by Surf3D, contains all three components of the magnetic field calculated. Now in order to compare the solution of the EFIE with the target function and the three scalar approaches, we choose the H_z-component of the calculated scattered field. This component does not correspond to the H_z-component derived in chapter 2 for the cylindrical waveguide. This field component is chosen, since it is proportional to the choice of $u^{(D)} \equiv \psi^{(2)}$ in the scalar approaches, see equation (2.52). Figures 4.18 and 4.19 shows the resulting scattered H_z^S-component in magnitude and phase. The result shows, as the other three approaches, the desired field distribution on the launcher wall. The magnitude of the result is much more agitated than the results of the scalar approaches. This is due to the fact, that the surface used for the evaluation of the scattered component is the actual perturbation surface, where

Parameter	Value
For the launcher domain:	
Number of triangles (geometry elements)	1532193
Number of field points	$1024 \times 1024 = 1048576$
Length of launcher domain [mm]	$0 \leq z \leq 350.15$
Vector correlation to Gaussian target function	$\eta_{vector} = 97.5\%$
Scalar correlation to Gaussian target function	$\eta_{scalar} = 98.3\%$
Time for computation	≈ 30 hours

Table 4.5: Main calculation parameters for the EFIE

as for the scalar approaches the evaluating surface is the straight waveguide with constant radius. The main calculation parameters for the EFIE are summarized in table 4.5. The calculation time necessary for the solution is about 30 hours, which is not astonishing considering the three-dimensional size of the problem.

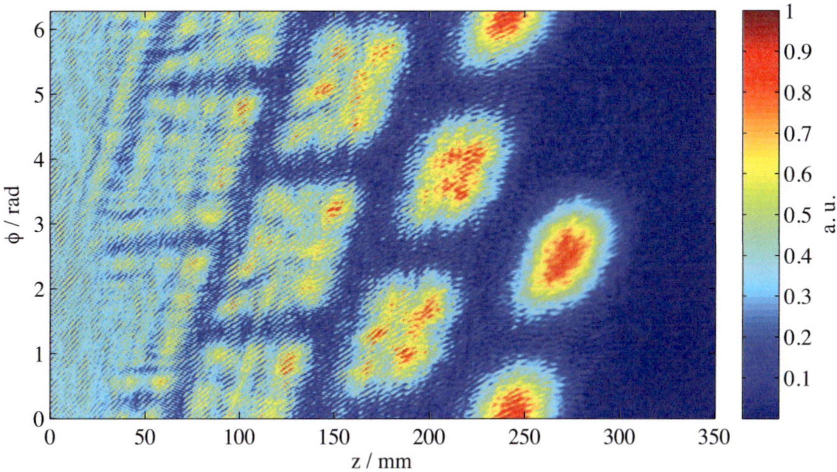

Figure 4.18: Magnitude (normalized) of the scattered field H_z^S resulting from the perturbed wall

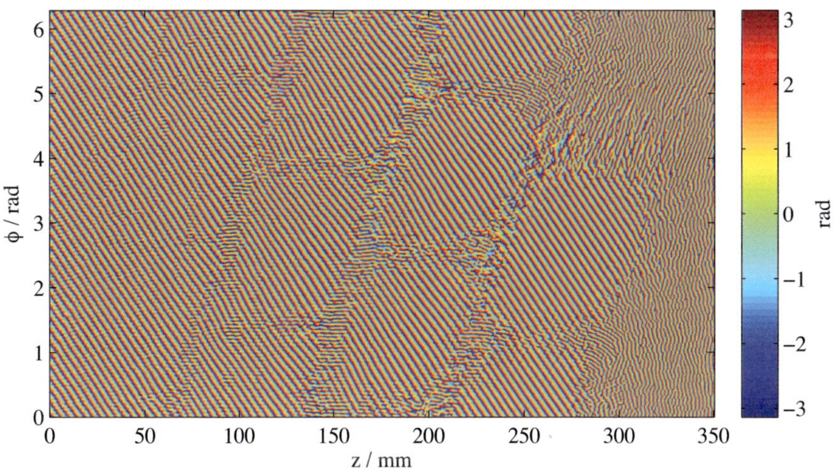

Figure 4.19: Phase of the scattered field H_z^S resulting from the perturbed wall

4.1.5 Comparison of the four approaches

For the comparison of the four different approaches, the vector correlation and scalar correlation coefficient, as defined in equations (2.80) and (2.79), are employed. The correlation is performed over the radiating aperture (the last Brillouin zone). Besides the correlations, we give the computational effort necessary for the final result in the comparison. Table 4.6 gives the different values of the correlations and the computational time effort including the corresponding CPU type.

Considering table 4.6 all four discussed methods are practically identical in terms of correlation coefficients (vector and scalar). Additionally, the correlation coefficients to the Gaussian target function are very alike. Considering the very high correlation of the CWD and DSD, an equivalence of the two methods is indicated. The best choice according to the comparison is the DSD, since the correlation to the Gaussian target function and to the EFIE have the highest values. The slightly lower correlation of the EFIE solution to the Gaussian target function could be due to the scattering of the rather agitated surface perturbation. The residual from section 2.3.5 for the DSD and the CWD is an indication for this fact, since $e_{20}^{r.m.s.} \approx 6\%$ and is rather high. As will be seen in the next section, the residual for the example launcher for the tapered algorithm using a smoothed perturbation surface is less than 2 % for the tapered $TE_{34,19}$-mode launcher. With the additional speed improvement, a future synthesis incorporating an iterative field calculation using the DSD should give the desired result in less time. Today these synthesis procedure take 1-2 days. Additionally, the speed enhancement gives the opportunity to investigate fabrication tolerances of the surface perturbation in an acceptable time frame. Furthermore, the electromagnetic field in launchers of multi-frequency gyrotrons with arbitrary surface perturbations can be calculated faster and eventually also synthesized.

Approach	PSD	DSD	CWD	EFIE
GTF	$\eta_v = 98.3\%$ $\eta_s = 98.7\%$	$\eta_v = 98.2\%$ $\eta_s = 98.7\%$	$\eta_v = 98.0\%$ $\eta_s = 98.7\%$	$\eta_v = 97.5\%$ $\eta_s = 98.3\%$
PSD		$\eta_v = 98.8\%$ $\eta_s = 99.4\%$	$\eta_v = 98.75\%$ $\eta_s = 99.39\%$	$\eta_v = 98.13\%$ $\eta_s = 99.13\%$
DSD			$\eta_v = 99.65\%$ $\eta_s = 99.83\%$	$\eta_v = 98.95\%$ $\eta_s = 99.46\%$
CWD				$\eta_v = 98.92\%$ $\eta_s = 99.44\%$
EFIE				
Comp. time	5.5 minutes	60 seconds	60 seconds	≈ 30 hours
CPU type	Core™ i5 @ 2.80GHz	Core™ i5 @ 2.80GHz	Core™ i5 @ 2.80GHz	Xeon® @ 2.40GHz

GTF:	Gaussian Target Function	CWD:	Cylindrical Wave Decomposition
PSD:	Point Source Diffraction	EFIE:	Electric Field Integral Equation
DSD:	Dipole Source Diffraction	Comp. time:	Computation Time
η_v:	η_{vector}	η_s:	η_{scalar}

Table 4.6: Comparison of the four discussed approaches for the $TE_{34,19}$-mode launcher

4.2 Launcher with tapered average radius

In the case of the calculation of the field for a launcher with a tapered average radius, the algorithm described in section 3.5 is employed. For the verification of the algorithm, the field of the example launcher from section 2.2.1 is calculated first with the coupled mode equations and then with the tapered algorithm for the DSD. For easier comparison of the data, an in-house coupled mode equations code, instead of the commercial code LOT, is used [Pre03]. For the comparison to the coupled mode equations the potential $\psi = \psi^{(2)} + \psi^{(1)}$ is used, since it is proportional to the H_z-component inside the waveguide. $\psi^{(2)}$ is used as before, when comparing to the EFIE. The second example is the comparison of the results for the $TE_{34,19}$-mode launcher with a mapped perturbation geometry to incorporate the taper [JTP$^+$09].

4.2.1 $TE_{22,6}$-mode 118 GHz launcher

The surface perturbation for the $TE_{22,6}$-mode launcher is depicted in figure 4.20. The periodic structure is very well illustrated. The magnitude of the resulting field ψ, using the DSD method, is presented in figure 4.21. The phase of the corresponding result is shown in figure 4.22. The results for H_z inside the waveguide, using the coupled mode equations, is shown in figure 4.23 and 4.24 in magnitude and phase respectively. Both results are in very good agreement. The correlation coefficients both vector and scalar for ψ and H_z, over the whole waveguide wall[2], are

$$\begin{aligned}
\eta_{vector} &= 96.88\% \quad \text{and} \\
\eta_{scalar} &= 99.9\% \,.
\end{aligned}$$

A very important issue, when comparing tapered launchers, for a good vector correlation is the exact same discretization, since a slight offset error in the discretization can give a vector correlation as low as $\eta_{vector} = 0.3\%$, while the scalar correlation η_{scalar} is still above 90%. This issue makes the comparison in tapered geometries rather cumbersome. This example shows the verification of the weighted overlap add algorithm for tapered launchers. One further issue using

---

[2]The coupled modes equations calculate the fields inside the closed waveguide, therefore the comparison is performed over the whole launcher wall

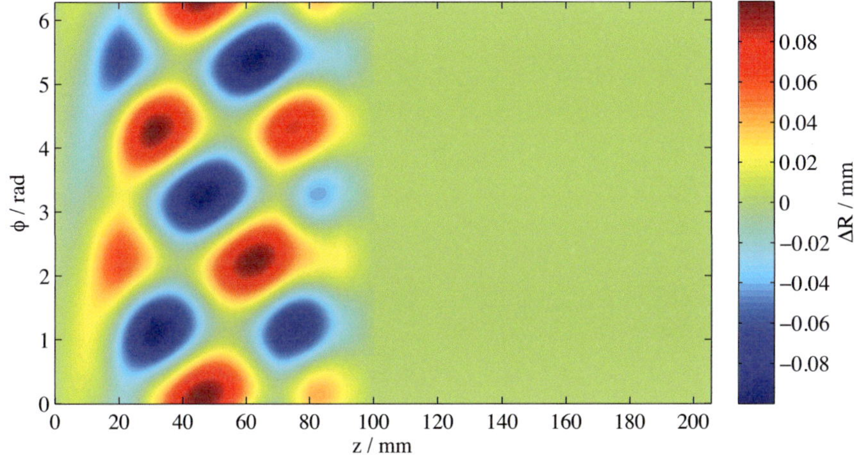

Figure 4.20: Perturbation according to (2.75) and Fig. 2.6 for the $TE_{22,6}$-mode launcher

the EFIE, is the evaluation of the integral for the calculation of the H_z-component inside the launcher for tapered geometries. The evaluation can give incorrect results near or on the waveguide wall [Nei11]. This is illustrated in figure 4.25, showing the magnitude of the scattered H_z^S on the waveguide wall for the tapered $TE_{22,6}$-mode 118 GHz launcher. This result differs drastically from the results obtained with the coupled mode equations and DSD.

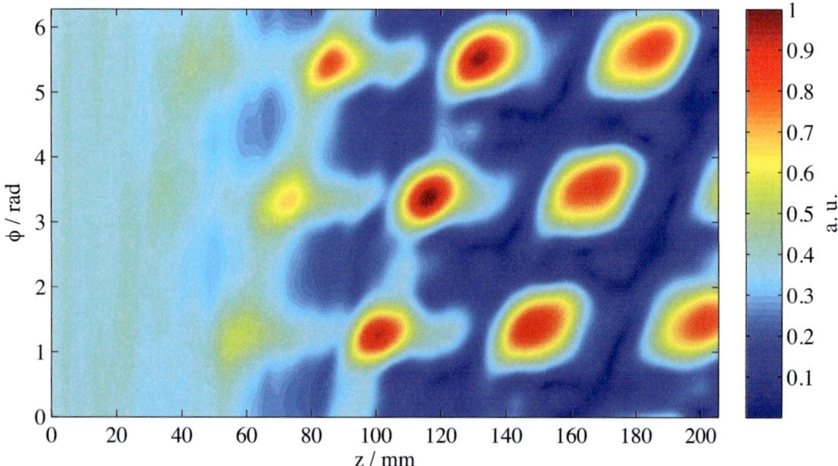

Figure 4.21: Magnitude (normalized) of ψ for the $TE_{22,6}$-mode 118 GHz launcher on the waveguide wall using DSD

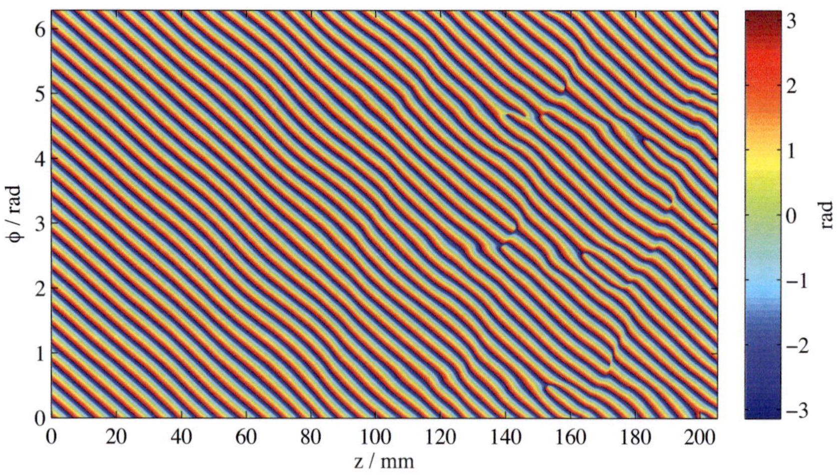

Figure 4.22: Phase of ψ for the $TE_{22,6}$-mode 118 GHz launcher on the waveguide wall using DSD

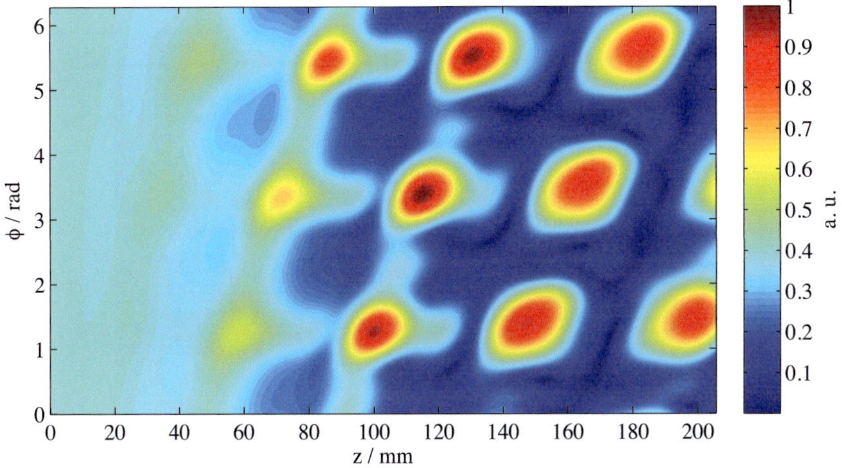

Figure 4.23: Magnitude (normalized) of H_z for the $TE_{22,6}$-mode 118 GHz launcher on the waveguide wall using coupled mode equations

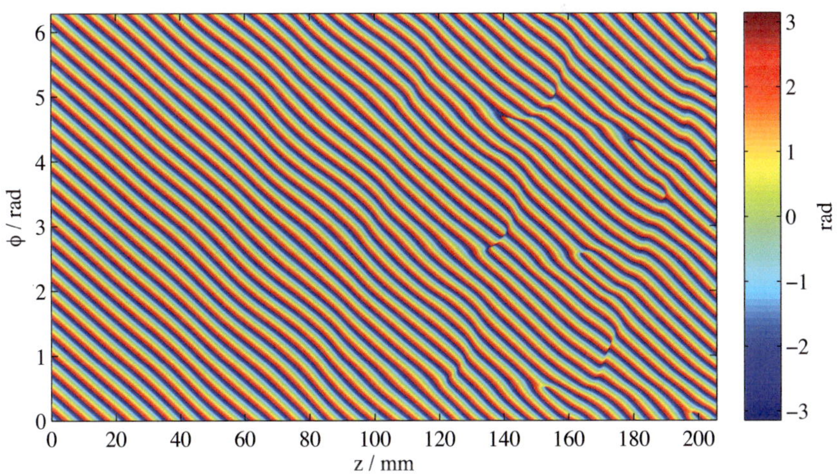

Figure 4.24: Phase of H_z for the $TE_{22,6}$-mode 118 GHz launcher on the waveguide wall using coupled mode equations

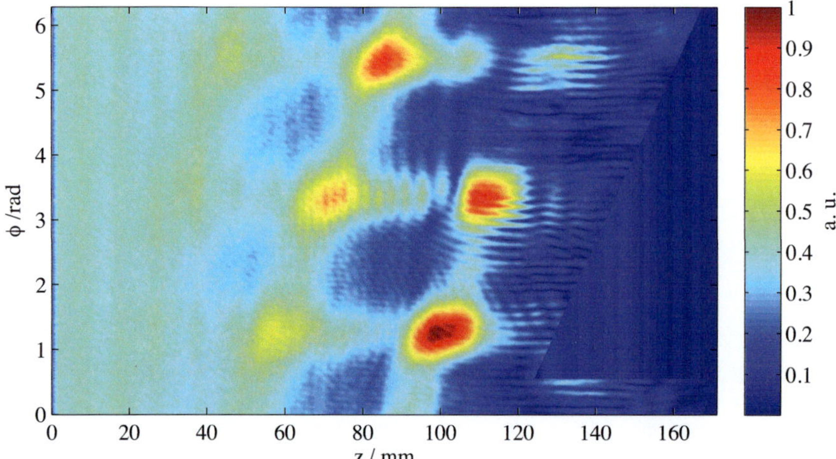

Figure 4.25: Incorrect result of the magnitude of the scattered H_z^S (normalized) on the waveguide wall for the $TE_{22,6}$-mode launcher using EFIE

4.2.2 $TE_{34,19}$-mode 170 GHz launcher

As a second example for the calculation of tapered launchers, the $TE_{34,19}$-mode launcher from section 4.1 is chosen. In order for the perturbation to incorporate the taper, the introduced mapping algorithm from [JTP$^+$09] is employed. Due to this mapping algorithm a non-equidistant meshing in z-direction is introduced. It becomes necessary to interpolate the perturbation onto the equidistant grid used in the DSD calculation. The wall perturbation was originally obtained by Jin [Jin11] and the interpolated perturbation is depicted in figure 4.26. In contrast to the perturbation from the un-tapered version, this surface perturbation has been smoothed (filtered) after the synthesis to produce less scattering. The final result of the calculation of the DSD is shown in figures 4.27 and 4.28 in magnitude and phase, respectively. The corresponding result H_z^S for the EFIE is shown in figures 4.29 and 4.30. The effects discussed earlier for the evaluation of the integral using

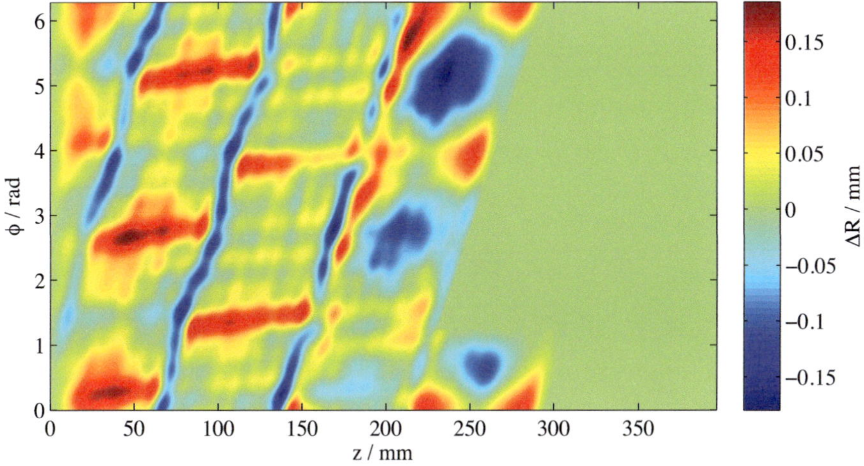

Figure 4.26: Interpolated wall perturbation of the tapered $TE_{34,19}$-mode launcher

the EFIE can be identified in figure 4.29 on the metal surface. In figure 4.30 these effects appear as phase jumps. These effects do not appear in and do not impact the radiated part of the field. The radiating aperture is given in figure 4.31. The vector and scalar correlation for the DSD and the EFIE, over the radiating aperture, are

$$\eta_{vector} = 98.9\% \quad \text{and}$$
$$\eta_{scalar} = 99.9\%.$$

The residual in this calculation is $e_{20}^{r.m.s.} \approx 1.9\%$. This gives an excellent agreement of the radiated fields calculated with the DSD and the EFIE. The computational effort for the two compared methods is the same as for the corresponding un-tapered algorithm. Up to now only the time and memory consuming method solving the EFIE gave the possibility to calculate tapered launchers with adaptive surface perturbation. Considering the enormous difference in computation time this gives for the first time, the possibility to calculate launchers with

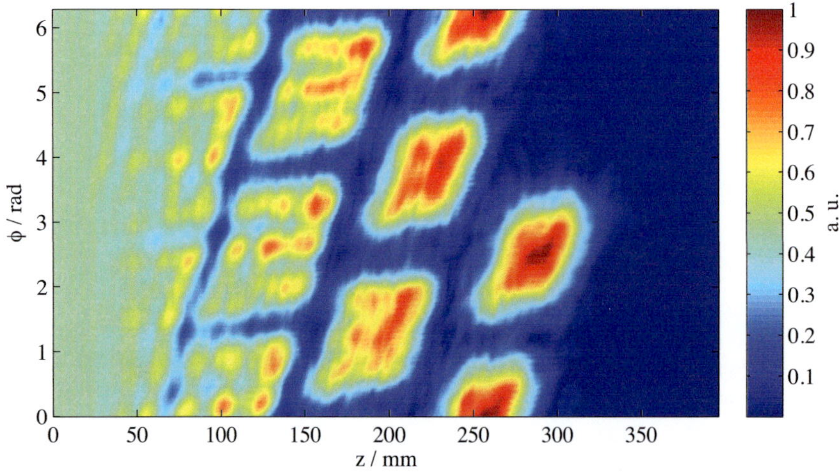

Figure 4.27: Magnitude of $u^{(D)}$ on the waveguide wall using DSD

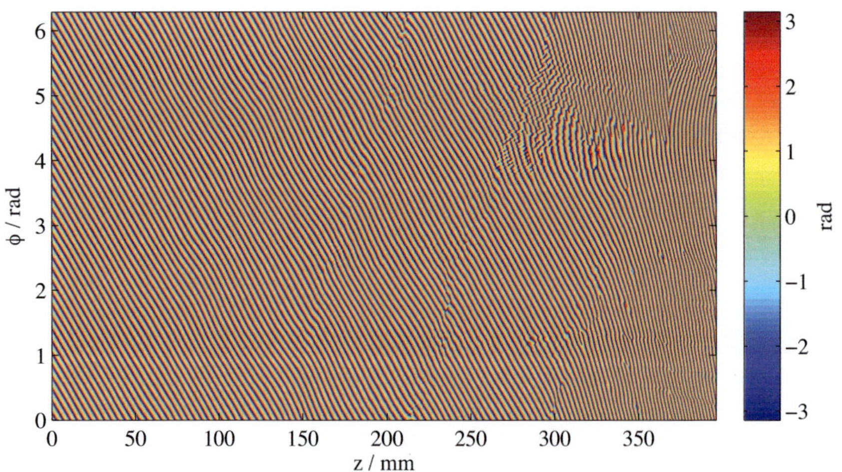

Figure 4.28: Phase of $u^{(D)}$ on the waveguide wall using DSD

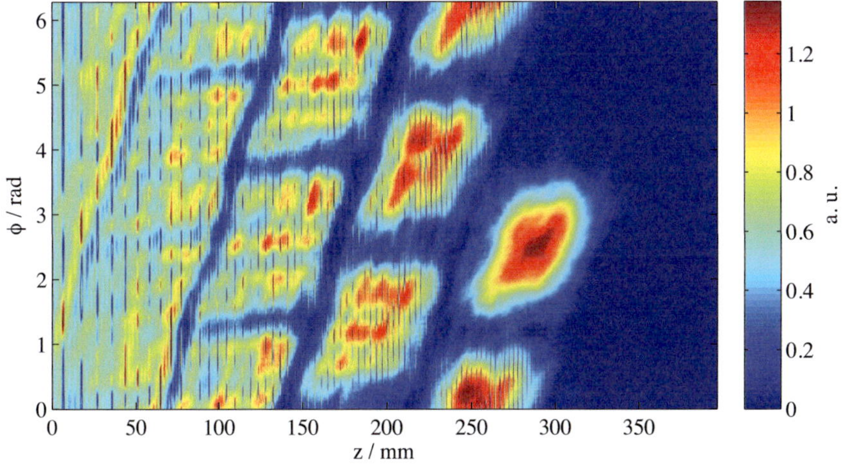

Figure 4.29: Magnitude of H_z^S on the waveguide wall using the EFIE

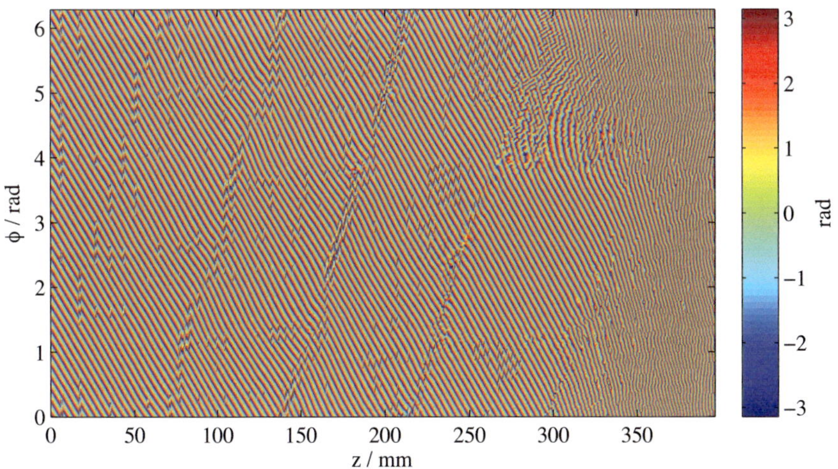

Figure 4.30: Phase of H_z^S on the waveguide wall using the EFIE

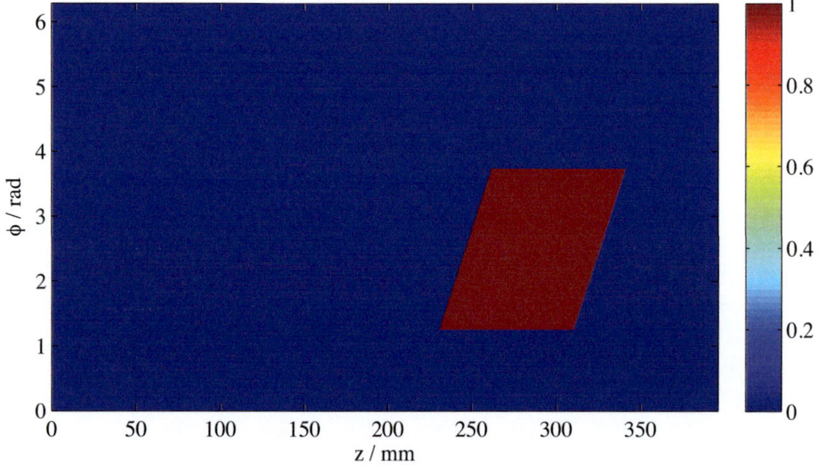

Figure 4.31: Radiating aperture

tapered average radius and arbitrary wall perturbation in a time frame acceptable for launcher design. This is possible due to the new introduction of the weighted overlap-add segmented convolution algorithm. The PSD implementation does not give the possibility to calculate tapered average radius launchers. Additionally, this enhancement gives the opportunity for the first time for the implementation of a future synthesis algorithm by iterative field calculation including tapered radii.

5 Conclusions and outlook

In this work, the diffraction and scattering in launchers of quasi-optical mode converters of high power gyrotrons has been investigated. Goal of the investigation was the comparison and classification of four different methods describing the wave propagation inside the launchers with adaptive surface perturbation. Of these four methods three are scalar quasi-optical methods, namely the Point Source Diffraction (PSD), the Dipole Source Diffraction (DSD) and the Cylindrical Wave Decomposition (CWD). The DSD and the CWD have been implemented in a computer program with a computational speed increase over the existing implementation of the PSD (TWLDO), of a factor of five. This enhancement is due to the newly introduced overlap-add segmented convolution algorithm and the reduction in problem size considering the diffraction (chapter 3). The fourth method solves the Electric Field Integral Equation (EFIE) implemented in the commercial full-wave code Surf3D. The EFIE solution is used as reference solution for the scalar approaches. All four methods have been theoretically described, their numerical aspects and algorithms have been discussed and calculation results of existing un-tapered launcher designs have been presented. The figures of merit, employed for the comparison, are the vector and scalar correlation coefficients given in chapter 2. The figures of merit are summarized in table 4.6. All four methods have a vector correlation $\eta_{vector} > 98\,\%$ and a scalar correlation coefficient $\eta_{vector} > 99\,\%$. This leads to the conclusion that all four methods are practically identical, considering the figures of merit. For the reasons mentioned in chapter 4, the calculation of the wave propagation in launchers is to be performed employing the DSD, for the computational speed enhancement. The speed improvement provides the possibility to investigate fabrication tolerances on the adaptive surface perturbation, as well as to conduct analysis of multi-frequency gyrotron launcher with adaptive surface perturbation. In addition, the correlation to the Gaussian target function, initially used for the synthesis, to the four methods is performed and the results reside in the same regime as for the comparison between the methods (Table 4.6). This again verifies the synthesis algorithm introduced in [JTP$^+$09].

The new extension of the DSD with the weighted overlap-add segmented convolution algorithm demonstrates the calculation of wave propagation in tapered average radius launchers with adaptive surface perturbation for the first time. The verification of the algorithm has also been performed by comparison to the full-wave solution of the commercial code, implementing the EFIE. The employed launcher surface design is a smoothed (filtered) surface perturbation synthesized with the PSD. To incorporate the taper, in an initially un-tapered design, a mapping algorithm [JTP$^+$09] was used, since the implementation of the PSD does not allow the calculation in tapered launchers. The figures of merit in the case for the tapered launcher calculation, comparing the DSD and the EFIE are $\eta_{vector} = 98.9\,\%$ and $\eta_{scalar} = 99.9\,\%$. This excellent agreement of a scalar and full-wave solution in combination with a drastic reduction of the computational time opens the possibility to future tapered launcher synthesis in one step with less stray radiation and diffraction. This result verifies the mapping algorithm used to obtain the tapered geometry. The reduction of the factor for computational time, going from EFIE solution to DSD, is in the order of 1800.

As a result of this work calculation of tapered average radius launchers with adaptive surface perturbations is made possible for the first time. In addition, based on this work, the possibilities of investigations on fabrication tolerances for launchers with adaptive surface perturbations, as well as calculation of launchers of multi-frequency gyrotrons with adaptive surface perturbations in a reasonable time frame are given. These steps lead further towards 2 MW class CW operating gyrotrons and high power CW multi-frequency gyrotrons.

Future improvements of the introduced algorithms and methods could include a conjugate-gradient algorithm instead of the employed successive approximation (Born's approximation) for faster convergence of the series solution. Furthermore the program could be enhanced with a multi-frequency capability to calculate launchers of multi-frequency gyrotrons. The employed Fast Fourier Transform (FFT) algorithm (FFTW) provides the possibility to incorporate a parallelization of the FFT, leading to a further speed improvement for launchers of multi-frequency gyrotrons. A major improvement would be the incorporation of the DSD method into a synthesis algorithm, in order to create tapered adaptive surface perturbation in one step, without mapping the surface as currently accomplished. To numerically estimate the stray radiation produced by the adaptive surface perturbation, the introduced depolarization factor from appendix A.2 could be implemented. Therefore a third figure of merit can be employed, in addition to the vector and scalar correlation coefficients, for determination of the quality of the surface perturbation and for the determination of diffraction and scattering in launchers of quasi-optical mode converters in gyrotrons. With these future improvements the classification and synthesis of tapered adaptive surface perturbation launchers for CW operating high power gyrotrons is possible.

Bibliography

[AS64]   Milton Abramowitz and Irene A. Stegun, editors. *Handbook of Mathematical Functions with Formulas, Graphs, and Mathematical Tables*. Dover, New York, ninth dover printing, tenth GPO printing edition, 1964.

[Bal89]   Constantine A. Balanis. *Advanced engineering electromagnetics*. Wiley, 1989.

[Bam89]   Richard Bamler. *Mehrdimensionale lineare Systeme : Fourier-Transformation und δ-Funktionen*. Nachrichtentechnik ; 20. Springer, Berlin, 1989.

[BC53]   Bevan B. Baker and Edward T. Copson. *The mathematical theory of Huygens principle*. Clarendon Pr., Oxford, 2. ed., reprint. edition, 1953.

[BD04]   A. A. Bogdashov and G. G. Denisov. Asymptotic theory of high-efficiency converters of higher-order waveguide modes into eigenwaves of open mirror lines. *Radiophysics and Quantum Electronics*, 47(4):283–295, 2004.

[Ber11]   Matthias Hermann Beringer. *Design studies towards a 4 MW 170 GHz coaxial-cavity gyrotron*. PhD thesis, Karlsruhe Institute of Technology, Karlsruhe, 2011. KIT Scientific Publishing, ISBN 978-3-86644-663-2.

[Bri88]   Elbert Oran Brigham. *The fast Fourier transform and its applications*. Prentice-Hall signal processing series. Prentice-Hall, Englewood Cliffs, N.J., 1988.

[Bro05]   Il'ja N. Bronstejn. *Taschenbuch der Mathematik*. Deutsch, 6., vollst. überarb. und erg. Aufl. edition, 2005.

[BS63]   Petr Beckmann and André Spizzichino. *The scattering of electromagnetic waves from rough surfaces*. International series of monographs on electromagnetic waves ; 4. Pergamon Pr., Oxford [u.a.], 1963.

[BW09]   Max Born and Emil Wolf. *Principles of Optics*. Cambridge, 5th printing, 7th expanded edition, 2009.

[Cát95]   M.F. Cátedra. *The CG-FFT method: application of signal processing techniques to electromagnetics*. Artech House antenna library. Artech House, 1995.

[CDK$^+$06] A. Chirkov, G. Denisov, M. Kulygin, V. Malygin, S. Malygin, A. Pavel'ev, and E. Soluyanova. Use of huygens' principle for analysis and synthesis of the fields in oversized waveguides. *Radiophysics and Quantum Electronics*, 49:344–353, 2006. 10.1007/s11141-006-0067-4.

[Che01]   W. Chew. *Fast and Efficient Algorithms im Computational Electromagnetics*. Artech House, 2001.

[Doa85]   J. L. Doane. Propagation and mode coupling in corrugated and smooth-wall circular waveguides. In K. J. Button, editor, *Infrared and Millimeter Waves*, volume 13, pages 123–170. Academic Press, 1985.

[DPV90]   G. G. Denisov, M. I. Petelin, and D. V. Vinogradov. Converter of high-mode of a circular waveguide into the main mode of a mirror line. *PCT Gazette*, 16:47–49, 1990.

[Dru02]   Oliver Drumm. *Numerische Optimierung eines quasi-optischen Wellentypwandlers für ein frequenzdurchstimmbares Gyrotron*. PhD thesis, Universität Karlsruhe, Juni 2002. Wissenschaftliche Berichte des Forschungszentrums Karlsruhe FZKA 6754.

[Edg93]   Christopher J. Edgcombe, editor. *Gyrotron oscillators: their principles and practice*. Taylor & Francis, 1993.

[FJT11]   Jens Flamm, Jianbo Jin, and Manfred Thumm. Wave propagation in advanced gyrotron output couplers. *Journal of Infrared, Millimeter and Terahertz Waves*, 32:887–896, 2011. 10.1007/s10762-011-9800-y.

[Gap59]   A.V. Gaponov. Interaction between electron fluxes and electromagnetic waves in waveguides. *Izv. Vyssh. Uchebn. Zaved. Radiofizika*, 2:450–462, 1959.

[Gol98]   Paul F. Goldsmith. *Quasioptical systems*. IEEE Press, 1998.

[Goo05]   Joseph Wilfred Goodman. *Introduction to Fourier optics*. Roberts, Englewood, Colo., 3. ed. edition, 2005.

[GR94]   I.S. Gradshteyn and I.M. Ryzhik. *Table of integrals, series, and products*, volume main. Acad. Press, New York, 5. ed. edition, 1994.

[Har61]   Roger F. Harrington. *Time-harmonic electromagnetic fields*. McGraw-Hill, 1961.

[Har68]   Roger F. Harrington. *Field computation by moment methods*. MacMillan, 1968.

[Har78]   F.J. Harris. On the use of windows for harmonic analysis with the discrete Fourier transform. *Proceedings of the IEEE*, 66(1):51 – 83, Jan. 1978.

[Jel00]   John Jelonnek. *Untersuchung des Lastverhaltens von Gyrotrons*. PhD thesis, Technische Universität Hamburg-Harburg, Düsseldorf, 2000.

[Jin07]   Jianbo Jin. Quasi-optical mode converter for a coaxial cavity gyrotron. Technical report, Forschungszentrum Karlsruhe, Karlsruhe, 2007. Wissenschaftliche Berichte des Forschungszentrum Karlsruhe FZKA 7264.

[Jin10]   Jianbo Jin. private communication, 2010.

[Jin11]   Jianbo Jin. private communication, 2011.

[JTP$^+$09]   Jianbo Jin, M. Thumm, B. Piosczyk, S. Kern, J. Flamm, and T. Rzesnicki. Novel numerical method for the analysis and synthesis of the fields in highly oversized waveguide mode converters. *IEEE Trans. on Microw. Theory Tech.*, 57(7):1661–1668, 2009.

[KBT04]   Machavaram V. Kartikeyan, Edith Borie, and Manfred K. A. Thumm. *Gyrotrons : high power microwave and millimeter wave technology.* Advanced texts in physics. Springer, Berlin, 2004.

[Ker96]   Stefan Kern. *Numerische Simulation der Gyrotron-Wechselwirkung in koaxialen Resonatoren.* PhD thesis, Universität Karlsruhe, Juli 1996. Wissenschaftliche Berichte des Forschungszentrums Karlsruhe FZKA 5837.

[KK06]   Karl-Dirk Kammeyer and Kristian Kroschel. *Digitale Signalverarbeitung : Filterung und Spektralanalyse; mit MATLAB-Übungen; mit 33 Tab.* Lehrbuch Elektrotechnik. Teubner, Stuttgart, 6., korr. und erg. Aufl. edition, 2006.

[Kuz97]   S. K. Kuzikov. Paraxial approach to description of wave propagation in irregular oversized waveguides. *Int. Journal of Infrared and Millimeter Waves*, 18(5):1001–1014, 1997.

[LV06]   S. Liao and R. J. Vernon. On the image approximation for electromagnetic wave propagation and pec scattering in cylindrical harmonics. *Progress In Electromagnetics Research*, 66:65–88, 2006. 10.2528/PIER06083002.

[MF53a]   P.M.C. Morse and H. Feshbach. *Methods of theoretical physics: Part 1.* International series in pure and applied physics. McGraw-Hill, 1953.

[MF53b]   P.M.C. Morse and H. Feshbach. *Methods of theoretical physics: Part 2.* International series in pure and applied physics. McGraw-Hill, 1953.

[Mic98]   Georg Michel. *Feldprofilanalyse und -synthese im Millimeterwellenbereich.* PhD thesis, Universität Karlsruhe, November 1998. Wissenschaftliche Berichte des Forschungszentrums Karlsruhe FZKA 6216.

[Nei04]   J. Neilson. Surf3d and LOT : computer codes for design and analysis of high-performance qo launchers in gyrotrons. In *Infrared and*

*Millimeter Waves, 2004 and 12th International Conference on Ter-
ahertz Electronics, 2004. Conference Digest of the 2004 Joint 29th
International Conference on*, pages 667 – 668, sept.-1 oct. 2004.

[Nei11]   Jeff Neilson. private communication, 2011.

[NS01]   R. Nevels and Chang-Seok Shin. Lorenz, Lorentz, and the gauge.
            Antennas and Propagation Magazine, IEEE, 43(3):70 –71, jun 2001.

[Nus04]   Gregory S. Nusinovic. *Introduction to the physics of gyrotrons*. John
            Hopkins studies in applied physics. Johns Hopkins Univ. Press, Balti-
            more, 2004.

[PADT07]   H.O. Prinz, A. Arnold, G. Dammertz, and M. Thumm. Analysis of a
            $TE_{22,6}$ 118-GHz Quasi-Optical Mode Converter. *Microwave Theory
            and Techniques, IEEE Transactions on*, 55(8):1697 –1703, aug. 2007.

[Pre03]   Julius Pretterebner. *Kompakte quasi-optische Antennen im überdi-
            mensionierten Rundhohlleiter*. PhD thesis, Universität Stuttgart, Juni
            2003.

[Pri07]   Oliver Prinz. *Dreidimensionale Feldberechnung für elektrisch große
            Geometrien und ihre Anwendung für multifrequente Hochleistungs-
            gyrotrons*. PhD thesis, Universität Karlsruhe, Oktober 2007. Wis-
            senschaftliche Berichte des Forschungszentrums Karlsruhe FZKA
            7351.

[PRM98]   Andrew F. Peterson, Scott L. Ray, and Raj Mittra. *Computational
            methods for electromagnetics*. IEEE/OUP series on electromagnetic
            wave theory. IEEE Press, New York, NY, 1998.

[PV07]   M. P. Perkins and R.J. Vernon. Mirror design for use in gyrotron
            quasi- optical mode converters. *IEEE Trans. on Plasma Science*,
            35(6):1747–1757, 2007.

[RS78]   Lawrence R. Rabiner and Ronald W. Schafer. *Digital processing of
            speech signals*. Prentice-Hall signal processing series. Prentice-Hall,
            Englewood Cliffs, NJ, 1978.

[Rze07]   Tomasz Rześnicki. *Analyse eines neuartigen 1.5 MW, 170 GHz Prototyp-Gyrotrons mit koaxialem Resonator.* PhD thesis, Universität Karlsruhe, Karlsruhe, 2007. Wissenschaftliche Berichte des Forschungszentrum Karlsruhe FZKA 7299.

[Sch59]   J. Schneider. Stimulated emission of radiation by relativistic electrons in a magnetic field. *Physical Review Letters*, 2:504–505, 1959.

[She67]   G. C. Shermann. Application of the convolution theorem to Rayleigh's integral formulas. *Journal of Opt. Soc. Am.*, 57:546–547, 1967.

[Sil86]   Samuel Silver, editor. *Microwave antenna theory and design.* IEE electromagnetic waves series ; 19. Peregrinus, London, reprinted edition, 1986.

[Smi11]   Julius O. Smith. *Spectral Audio Signal Processing, October 2008 Draft.* `http://ccrma.stanford.edu/~jos/sasp/`, accessed 2011. online book.

[Sol59]   L. Solymar. Spurious mode generation in nonuniform waveguides. *IRE Transactions on Microwave Theory and Techniques*, 7(3):379–383, 1959.

[Som89]   Arnold Sommerfeld. *Vorlesungen über theoretische Physik*, volume 4: Optik. Verlag Harri Deutsch, Leipzig, nachdr. d. 3., durchges. aufl. edition, 1989.

[Som92]   Arnold Sommerfeld. *Vorlesungen über theoretische Physik*, volume 6: Partielle Differentialgleichungen der Physik. Verlag Harri Deutsch, Thun, nachdr. d. 6., aufl. edition, 1992.

[Thu84]   M. Thumm. High power millimeter-wave mode converters in overmoded circular waveguides using periodic wall perturbations. *International Journal of Electronics*, 57:1225–1246, 1984.

[Thu97]   M. Thumm. Modes and mode conversion in microwave devices. In R.A. Cairns and A.D.R. Phelps, editors, *Generation and application of high power microwaves*, pages 121–171. SUSSP Publications and Institute of Physics Publishing, 1997.

[Thu11]   Manfred Thumm. *State-of-the-Art of High Power Gyro-Devices and Free Electron Masers Update 2010*. KIT Publishing, KIT Scientific Reports 7575, 2011.

[Twi58]   R.Q. Twiss. Radiationtransfer and the possibility of negative absorption in radio astronomy. *Australian Journal of Physics*, 11:564–579, 1958.

[vB07]   Jean van Bladel. *Electromagnetic fields*. IEEE Press-Wiley, Hoboken, N.J., 2. ed. edition, 2007.

[VZP75]   S.N. Vlasov, L.I. Zagyradskaja, and M.I. Petelin. Transformation of a whispering gallery mode, propagating in a circular waveguide, into a beam of waves. *Radio Engineering and Electron Physics*, 20:14–17, 1975.

[Wei69]   L.A. Weinstein. *Open Resonators and Open Waveguides*. Golem Press, 1969.

[Wey19]   H. Weyl. Ausbreitung elektromagnetischer Wellen über einem ebenen Leiter. *Annalen der Physik*, 365(21):481–500, 1919.

[Wie95]   Andreas Wien. *Ein Beitrag zur Analyse von quasi-optischen Wellentypwandlern in Hochleistungsgyrotrons*. PhD thesis, Universität Karlsruhe, Juli 1995. Wissenschaftliche Berichte des Forschungszentrums Karlsruhe FZKA 5797.

A Appendix

A.1 Derivative of the Green's function in cylindrical coordinates

For the derivation of the Green's function in cylindrical coordinates we start from equation (2.90):

$$
\begin{aligned}
\nabla' G \cdot \vec{n}' &= \frac{1}{4\pi} \frac{d}{dr_0} \left(\frac{e^{-jkr_0}}{r_0} \right) \vec{n}' \cdot \frac{\vec{r}_0}{r_0} \\
&= -\frac{1}{4\pi} \left(jk + \frac{1}{r_0} \right) \cdot \frac{e^{-jkr_0}}{r_0} \vec{n}' \cdot \frac{\vec{r}_0}{r_0}
\end{aligned}
\tag{A.1}
$$

Equating this on a cylindrical surface of the launcher at $\rho = R_0$ with the following given substitutions

$$
\begin{aligned}
r_0 &= \sqrt{(z - z')^2 + (y - y')^2 + (x - x')^2} \\
x &= R_0 \cdot \cos\varphi \\
y &= R_0 \cdot \sin\varphi \\
z &= z \\
&\Longrightarrow \\
r_0 &= \sqrt{(z - z')^2 + 4 \cdot R_0^2 \cdot \sin^2\left(\frac{\varphi - \varphi'}{2} \right)} \\
\vec{n} &= \vec{e}_r
\end{aligned}
$$

$$\vec{n} \cdot \vec{r}_0 = \begin{pmatrix} \cos\varphi' \\ \sin\varphi' \\ 0 \end{pmatrix} \cdot \begin{pmatrix} R_0 \cdot (\cos\varphi - \cos\varphi') \\ R_0 \cdot (\sin\varphi - \sin\varphi') \\ z - z' \end{pmatrix}$$

$$\vec{n} \cdot \vec{r}_0 = -2 \cdot R_0 \cdot \sin^2 \left(\frac{\varphi - \varphi'}{2} \right)$$

$$\vec{n} \cdot \frac{\vec{r}_0}{r_0} = \frac{-2 \cdot R_0 \cdot \sin^2 \left(\frac{\varphi - \varphi'}{2} \right)}{\sqrt{(z - z')^2 + 4 \cdot R_0^2 \cdot \sin^2 \left(\frac{\varphi - \varphi'}{2} \right)}}$$

$$dS = R_0 \, d\varphi \, dz$$

gives

$$\nabla' G \cdot \vec{n}' = \frac{1}{4\pi} \left(jk + \frac{1}{r_0} \right) \cdot \frac{-2 \cdot R_0 \cdot \sin^2 \left(\frac{\varphi - \varphi'}{2} \right)}{r_0} \cdot \frac{e^{-jkr_0}}{r_0} \tag{A.2}$$

A.2 Depolarization factor for estimation of stray radiation

In order to estimate the stray radiation inside the launcher of a quasi-optical mode converter, the use of a depolarization factor is suggested. This depolarization factor is used to estimate the power transferred into the depolarized component (TM modes), due to the scattering at the rough surface perturbation. The derivation follows chapter 8 in [BS63]. The depolarization factor for any rough surface is given as (section 8.4 in [BS63]):

$$p_2 = \frac{E_2^+}{E_2^-} \tag{A.3}$$

where the "+"-superscript denotes parallel (TM) and "$-$"-superscript denotes perpendicular polarization (TE). The subscript 2 denotes the reflected wave from the rough surface. This depolarization factor p_2 can be expressed in terms of the geometrical characteristics of the rough surface. In general it is given as

$$p_2 = \frac{p_1 \left(R^- \tan\beta \tan\beta_2 + R^+ \right) - R^- \tan\beta + R^+ \tan\beta}{R^- + R^+ \tan\beta \tan\beta_2 - p_1 \left(R^- \tan\beta_2 - R^+ \tan\beta \right)} \tag{A.4}$$

where p_1 denotes the incoming polarization of the wave. R^- and R^+ are determined by the boundary conditions of the surface and have to be adapted, since the calculation of launchers employs $u \equiv \psi^{(1)} \equiv F_z^{(1)}$, more precisely the inward traveling part of it and not the electric field E. $\tan\beta$ and $\tan\beta_2$ are determined by the inclination between the surface and the wave. For TE-modes near cut-off $\beta_2 \approx \beta$ and since u is perpendicular $p_1 = 0$ as defined in [BS63], therefore equation A.4 reduces to:

$$p_2 = \frac{-R^- \tan\beta + R^+ \tan\beta}{R^- + R^+ \tan^2\beta} \tag{A.5}$$

Now determining the coefficients R^- and R^+ for u, by using section 4.2.1 in [Mic98], yields:

$$\begin{aligned} R_F^- &= 1 \\ R_F^+ &= -1 \end{aligned}$$

Substituting this result in A.5 gives the final result:

$$p_2 = \frac{-2\tan\beta}{1 - \tan^2\beta} = -\tan 2\beta \tag{A.6}$$

$\tan\beta$ is given from [BS63], after substitution of the parameters, as:

$$\tan\beta = \frac{\vec{e}_\varphi \cdot \nabla\left(\Delta R(\varphi, z)\right)}{\frac{m}{\chi'_{mn}} - \vec{e}_z \cdot \nabla\left(\Delta R(\varphi, z)\right) \cdot \frac{k_\rho}{k}\sqrt{1 - \left(\frac{m}{\chi'_{mn}}\right)^2}} \tag{A.7}$$

Substituting $u \equiv u^{(D)} \equiv \psi^{(1)}$ for E_2^- in equation A.3, squaring the whole equation and integrating over the whole launcher surface, the depolarized power can be estimated as:

$$P_{depol} \approx \iint\limits_S u^2(\varphi, z) \cdot \tan^2(2\beta)\, dS \tag{A.8}$$

For the iterative algorithms, this power has to be calculated in each iteration and summed up, to give an estimation of the total depolarized power $P_{total,depol}$. This derivation has to be verified by implementation in a computer program, but since a similar estimation is found in [CDK$^+$06], the estimation should give good results.

Acknowledgments

The present work was accomplished during my time as teaching assistant at the Institute of High Frequency Techniques and Electronics (IHE) and at the same time research assistant at the Institute for Pulsed Power and Microwave Technology (IHM) at Karlsruhe Institute of Technology (KIT).

At first my deepest gratitude goes to Prof. Dr. rer. nat. Dr. h.c. Manfred Thumm for the opportunity to study the fascinating field of high power gyrotrons and for being my doctoral adviser. His confidence in me and the freedom he gave me, helped me to accomplish this work.

I am also grateful to Prof. Dr.-Ing. Dr. h.c. Klaus Schünemann from TU Hamburg-Harburg for his great interest in my work and being my second reviewer.

I would like to thank all members of the High Power Microwave department at IHM for the excellent working environment, the inspiring discussions during coffee breaks and our extracurricular activities. Particularly, I would like to thank Dr. Rudolf Schneider for the innumerable discussions on solving diffraction problems and Dr. Stefan Kern for his valuable discussions and feedback. My thanks goes to Dr. Jianbo Jin for providing me the surface perturbations necessary for my work. Without this support, my work would have not been possible. I want to express my deepest appreciation to Dipl.-Ing. Andreas Schlaich for the great cooperation during his student project and diploma thesis and for our inspiring discussions about all different kinds of important things in life. I would also like to thank all the colleagues at IHE for the support in "teaching assistant matters".

Many thanks to Prof. Dr.-Ing. John Jelonnek for our discussions and his excellent guidance.

My very special thanks goes to family, especially to my mother, who was always there for me. Without her, I would not be, who I am today. And it goes to my father for his tireless support and for being my best role model. Furthermore thanks to my brother Philipp for reminding me of the "non-engineering" side of life.

Finally, I am deeply grateful to my wife Silke for her love and continuous support and for her sacrifice by giving me the time and space I needed to accomplish this work. I am lucky to have her by my side.

Karlsruhe, March 2012
Jens Flamm